D0909334

Grundlehren der mathematischen Wissenschaften 277

A Series of Comprehensive Studies in Mathematics

Grundlehren der mathematischen Wissenschaften

A Series of Comprehensive Studies in Mathematics

A Selection

190. Faith: Algebra: Rings, Modules, and Categories I
191. Faith: Algebra II, Ring Theory
192. Mal'cev: Algebraic Systems
193. Pólya/Szegö: Problems and Theorems in Analysis I
194. Igusa: Theta Functions
195. Berberian: Baer*-Rings
196. Athreya/Ney: Branching Processes
197. Benz: Vorlesungen über Geometrie der Algebren
198. Gaal: Linear Analysis and Representation Theory
199. Nitsche: Vorlesungen über Minimalflächen
200. Dold: Lectures on Algebraic Topology
201. Beck: Continuous Flows in the Plane
202. Schmetterer: Introduction to Mathematical Statistics
203. Schoeneberg: Elliptic Modular Functions
204. Popov: Hyperstability of Control Systems
205. Nikol'skii: Approximation of Functions of Several Variables and Imbedding Theorems
206. André: Homologie des Algèbres Commutatives
207. Donoghue: Monotone Matrix Functions and Analytic Continuation
208. Lacey: The Isometric Theory of Classical Banach Spaces
209. Ringel: Map Color Theorem
210. Gihman/Skorohod: The Theory of Stochastic Processes I
211. Comfort/Negrepontis: The Theory of Ultrafilters
212. Switzer: Algebraic Topology—Homotopy and Homology
213. Shafarevich: Basic Algebraic Geometry
214. van der Waerden: Group Theory and Quantum Mechanics
215. Schaefer: Banach Lattices and Positive Operators
216. Pólya/Szegö: Problems and Theorems in Analysis II
217. Stenström: Rings of Quotients
218. Gihman/Skorohod: The Theory of Stochastic Process II
219. Duvant/Lions: Inequalities in Mechanics and Physics
220. Kirillov: Elements of the Theory of Representations
221. Mumford: Algebraic Geometry I: Complex Projective Varieties
222. Lang: Introduction to Modular Forms
223. Bergh/Löfström: Interpolation Spaces. An Introduction
224. Gilbarg/Trudinger: Elliptic Partial Differential Equations of Second Order
225. Schütte: Proof Theory
226. Karoubi: K-Theory, An Introduction
227. Grauert/Remmert: Theorie der Steinschen Räume
228. Segal/Kunze: Integrals and Operators
229. Hasse: Number Theory
230. Klingenberg: Lectures on Closed Geodesics
231. Lang: Elliptic Curves: Diophantine Analysis
232. Gihman/Skorohod: The Theory of Stochastic Processes III
233. Stroock/Varadhan: Multi-dimensional Diffusion Processes
234. Aigner: Combinatorial Theory

Continued after Index

William Fulton
Serge Lang

Riemann-Roch Algebra

Springer-Verlag
New York Berlin Heidelberg Tokyo

William Fulton
Department of Mathematics
Brown University
Providence, RI 02912
U.S.A.

Serge Lang
Department of Mathematics
Yale University
New Haven, CT 06520
U.S.A.

AMS Subject Classification: 14C40

Library of Congress Cataloging in Publication Data
Fulton, William
Riemann–Roch algebra.
(Grundlehren der mathematischen Wissenschaften; 277)
Bibliography: p.
Includes index.
1. Geometry, Algebraic. 2. Riemann–Roch theorems.
I. Lang, Serge, 1927– . II. Title. III. Series.
QA564.F85 1985 512′.33 84-26842

Typeset by Composition House Ltd., Salisbury, England.
Printed and bound by R. R. Donnelley & Sons, Harrisonburg, Virginia.
Printed in the United States of America.

9 8 7 6 5 4 3 2 1

ISBN 0-387-96086-4 Springer-Verlag New York Berlin Heidelberg Tokyo
ISBN 3-540-96086-4 Springer-Verlag Berlin Heidelberg New York Tokyo

Contents

Introduction

In various contexts of topology, algebraic geometry, and algebra (e.g. group representations), one meets the following situation. One has two contravariant functors K and A from a certain category to the category of rings, and a natural transformation

$$\rho: K \to A$$

of contravariant functors. The Chern character being the central example, we call the homomorphisms

$$\rho_X: K(X) \to A(X)$$

characters. Given $f: X \to Y$, we denote the pull-back homomorphisms by

$$f^K: K(Y) \to K(X) \quad \text{and} \quad f^A: A(Y) \to A(X).$$

As functors to abelian groups, K and A may also be covariant, with push-forward homomorphisms

$$f_K: K(X) \to K(Y) \quad \text{and} \quad f_A: A(X) \to A(Y).$$

Usually these maps do not commute with the character, but there is an element $\tau_f \in A(X)$ such that the following diagram is commutative:

$$\begin{array}{ccc} K(X) & \xrightarrow{\tau_f \cdot \rho_X} & A(X) \\ {\scriptstyle f_K}\downarrow & & \downarrow{\scriptstyle f_A} \\ K(Y) & \xrightarrow[\rho_Y]{} & A(Y) \end{array}$$

The map in the top line is ρ_X multiplied by τ_f.

When such commutativity holds, we say that Riemann–Roch holds for f. This type of formulation was first given by Grothendieck, extending the work of Hirzebruch to such a relative, functorial setting. Since then several other theorems of this Riemann–Roch type have appeared. Underlying most of these there is a basic structure having to do only with elementary algebra, independent of the geometry. One purpose of this monograph is to describe this algebra independently of any context, so that it can serve axiomatically as the need arises.

A common feature of these Riemann–Roch theorems is that a given morphism f is factored into $p \circ i$:

$$X \xrightarrow{i} P \xrightarrow{p} Y,$$

where i is a closed imbedding and p is a bundle projection. One constructs a deformation from i to the zero-section imbedding of X in the normal bundle to X in P, suitably completed at infinity. General procedures, which we axiomatize here, allow one to deduce a general Riemann–Roch theorem from the elementary cases of imbeddings in and projections from bundles; these cases are usually handled by direct calculation.

We illustrate the formalism by giving a complete elementary account of Grothendieck's Riemann–Roch theorem in the context of schemes and local complete intersection morphisms, as first presented in [SGA 6]. Here $K(X)$ is the Grothendieck ring of locally free sheaves on X, and $A(X)$ is an associated graded group of $K(X)$, with rational coefficients. To prepare for this we include self-contained discussions of several important subjects from algebra and algebraic geometry, such as: λ-rings, Adams operations, γ-filtrations, Chern classes, algebraic K-theory, regular imbeddings and Koszul complexes, sheaves on projective bundles, and local complete intersections.

Manin's very useful notes [Man] were also written to give an accessible account of parts of [SGA 6], for the case of imbeddings of nonsingular varieties. Several developments since then allow us to give both a more elementary and more complete treatment, including a complete proof of the main theorem, as well as some conjectures left open in [SGA 6]. Most important among these developments are: (a) an understanding of deformation to the normal bundle (cf. [J], [BFM 1], [V], [BFM 2], [FM]); (b) the use of Castelnuovo–Mumford "regular" sheaves on projective bundles (cf. [Q]). Among the resulting improvements we mention:

(1) A proof that the γ-filtration on $K(X)$ is finer than the topological filtration (V, §3).
(2) A Riemann–Roch theorem for the Adams operations ψ^j without denominators (V, §6).
(3) An elementary construction of the push-forward f_K for a projective local complete intersection morphism f (V, §4).

Of these, (1) and (2) were conjectured in [SGA 6]. Other features included are:

(4) An Intersection Formula for K-theory (VI, §1).
(5) A direct proof, using a power-series calculation of R. Howe, for Grothendieck Riemann–Roch for bundle projections (II, §2).
(6) An equivalence between forms of Riemann–Roch for the Chern character and Adams operators (III, §4).

Chapter I contains an elementary treatment of λ-rings and Chern classes; the excellent exposition of Atiyah and Tall [AT] can be referred to for more on λ-rings. We include a proof of a splitting principle for abstract Chern classes; in our application in Chapter V, however, this splitting principle will be evident, so the reader can skip this proof.

In Chapter II we develop the abstract Riemann Roch formalism. The main new feature here is an axiomatic formulation of the deformation to the normal bundle: to prove a Riemann–Roch theorem for a given imbedding, it suffices to "deform" it to an "elementary imbedding" for which one knows the theorem. We also axiomatize the dual case of an "elementary projection".

Chapter III describes the γ-filtration of Grothendieck, and constructs Chern classes in the associated graded ring.

Chapter IV is a chapter of "intermediate algebraic geometry", which could supplement a text such as Hartshorne's [H]. We establish the basic category of algebraic geometry for which we shall prove the Riemann–Roch formula, namely the category of regular morphisms. By this we mean morphisms which can be factored into a local complete intersection imbedding, and a projection from a projective bundle. We include a short proof of Micali's theorem on regular sequences, and basic facts about regular imbeddings, conormal sheaves, and blowing up. Theorem 4.5 on the residual structure of a proper transform is, we believe, new. The culmination of this chapter is a simple construction of the deformation to the normal bundle. Many of the results of Chapter IV are not needed for the proof of Riemann–Roch proper, but are included for completeness.

All these ideas come together in Chapter V, where the λ-ring $K(X)$ is shown to satisfy the abstract properties of the first three chapters. The Grothendieck Riemann–Roch theorem (including the version without denominators), and analogous theorems for the Adams operators, follow quickly.

Chapter VI contains an Intersection Formula in the context of K-theory which seems to be new in this generality, and which is analogous to the "excess intersection formula" of [FM], see also [F 2], Theorem 6.3. The formula is proved by using the general formalism of basic deformations, together with the geometric construction of the deformation to the normal bundle. This follows a pattern similar to the proof for

Riemann–Roch itself, and provides another striking application of the formalism of Chapter II.

In Chapter VI, we also discuss the relation of the Grothendieck group of locally free sheaves with the Grothendieck group of all coherent sheaves. We give an application to the calculation of an exact sequence for K of a blow up of a regularly imbedded subscheme, relying on the Intersection Formula. Finally, we discuss briefly and incompletely how Riemann–Roch can be extended beyond the case of local complete intersections. In addition, we sketch several other contexts where the formalism developed here can be applied. It would take another book to give a systematic treatment of these topics, including the relations between K-theory, the Chow group and étale cohomology in a more schemy and sheafy context than [F 2].

We have made our exposition self-contained from [H] for algebraic geometry, [L] for general algebra, and the simpler parts of [Mat] for a little more commutative algebra. Thus we have included proofs of elementary facts whenever necessary to achieve this.

At least in first reading, the reader interested only in a fast proof of Riemann–Roch is advised to skim Chapters I, IV, and the first half of Chapter V. More is included in these chapters than is strictly needed for Riemann–Roch, with the hope that this important material will be more accessible than its previous position in SGA and EGA permit. Those interested primarily in the Riemann–Roch theorem should concentrate on Chapters II, III, and V.

We have not discussed applications to the theory of group representations. For these, we refer especially to the articles by Atiyah–Tall, Evens, Kahn, Knopfmacher, Thomas, as well as Grothendieck's general discussion as listed in the Bibliography. On the other hand, the applications to group representations are not independent of those to algebraic geometry. Even though the K-groups can be defined in terms of modules, one can analyze them via considerations of topology, classifying spaces, and algebraic geometry, so there is a considerable amount of feedback.

We also do not discuss applications to topology. We refer to the lectures by Atiyah [At] and Bott [Bo] for some K-theory like that of Chapters I and III in a topological context, stopping short of Riemann–Roch theorems, however.

We hope that the simpler logical structure of the proofs which emerges in this treatise will make it easier to understand these results, and to find new situations to which this "Riemann–Roch algebra" applies.

CHAPTER I

λ-Rings and Chern Classes

This chapter describes first the basic ring structure of the objects to be encountered later in a more geometric context. The algebra involved is elementary and self-contained. We have axiomatized certain notions which originally arose in the theory of vector bundles. Actually we work with two rings, one of them usually graded. We also develop the formalism of Hirzebruch polynomials, which belongs to the basic theory of symmetric functions. We have preserved original names like Chern classes, Todd character, etc., although the algebra involved here deals only with a pair of rings and some elementary formal manipulation of power series, independently of the geometry from which they came.

We now make additional comments concerning the way these notions arise in applications to algebraic geometry and group representations. These are not necessary for a logical understanding of the chapter. However, we may have at least two categories of readers: those who know some Riemann–Roch theory previously and are principally interested in a quick proof of Grothendieck Riemann–Roch; and those who have more limited knowledge in this direction and are thus directly interested in the more elementary material. Our additional comments are addressed to this second category.

A fundamental aim of algebraic geometry is to study divisor classes, or equivalently isomorphism classes of line bundles. More generally, one wishes to study vector bundles, with certain equivalence relations. The Grothendieck relations are those which to each short exact sequence

$$0 \to E' \to E \to E'' \to 0$$

gives the relation

$$[E] = [E'] + [E''].$$

The group of isomorphism classes of vector bundles over a space X modulo these relations is called the Grothendieck group $K(X)$. It has both covariant and contravariant functorial properties, although the covariant ones are much more subtle.

The addition is induced by the direct sum, and there is also a multiplication induced by the tensor product, so that $K(X)$ is in fact a ring. The class of the trivial line bundle is the unit element.

This ring has various structures. First, it has an augmentation, which to E associates its rank $\varepsilon(E)$. Then ε extends to an augmentation on $K(X)$ (algebra homomorphism into $\mathbf{Z}$). The vector bundles themselves generate a semigroup under addition. In §1, we axiomatize this structure by defining "positive elements" whose properties are modelled on those of vector bundles. The elements of augmentation 1 correspond to line bundles, and are thus called line elements.

Second, the ring $K(X)$ has another operation induced by the alternating product. To each integer $i \geqq 0$ we have $\Lambda^i E$, and therefore its class $[\Lambda^i E]$ denoted by $\lambda^i(E)$. A standard elementary formula for the direct sum $E = E' \oplus E''$ of free modules reads

$$\Lambda^n(E) \approx \bigoplus_{i=0}^{n} (\Lambda^i E' \otimes \Lambda^{n-i} E'').$$

Passing to the classes in the K-group, we get the relation

$$\lambda^n(x + y) = \sum_{i=0}^{n} \lambda^i(x)\lambda^{n-i}(y).$$

But this relation amounts to saying that the map

$$x \mapsto \sum \lambda^i(x)t^i = \lambda_t(x) \qquad \text{by definition}$$

is a homomorphism from the additive group of $K(X)$ to the multiplicative group of power series with constant term equal to 1. This gives rise to the notion of λ-ring. A great deal of the formalism of Riemann–Roch algebra can be developed for the general λ-rings. The reader should read simultaneously the beginning of Chapter I and the beginning of Chapter V to see the parallelism between the abstract algebra and the geometric construction giving rise to this algebra.

In the theory of group representations, one may start with the category of finite-dimensional vector spaces over a field k, and a representation of a (finite) group G on the space. Then again we have direct sums, tensor products of (G, k)-spaces and the analogous definition of λ-ring, formed by the isomorphism classes of such spaces modulo the relations in the Grothendieck group. The positive elements are just the classes of such spaces as distinguished from the group generated by them in the Grothendieck group.

In §2 we shall discuss a particular extension of a λ-ring, which gives an axiomatization for the extension obtained from a projective bundle.

The corresponding geometric case is discussed in Chapter V, Theorem 2.3 and Corollary 2.4. Since the existence of the extension is proved in a self-contained way by geometric means in Chapter V, the reader interested only in the geometric application can omit the existence proof of Theorem 2.1 in this chapter. The corresponding graded extension will be constructed in §3.

I §1. λ-Rings with Positive Structure

Let K be a commutative ring. For each integer $i \geqq 0$ suppose given a mapping

$$\lambda^i: K \to K$$

such that $\lambda^0(x) = 1$, $\lambda^1(x) = x$ for all $x \in K$, and if we put

$$\lambda_t(x) = \sum \lambda^i(x)t^i$$

then the map

$$x \mapsto \lambda_t(x)$$

is a homomorphism. This condition is equivalent with the conditions

$$\lambda^k(x + y) = \sum_{i=0}^{k} \lambda^i(x)\lambda^{k-i}(y) \tag{1.1}$$

for all positive integers k. A ring with such a family of maps λ^i is called a **λ-ring**.

In addition, we suppose that the λ-ring has what we shall call a **positive structure**. By this we mean:

A surjective ring homomorphism

$$\varepsilon: K \to \mathbf{Z}$$

called the **augmentation**.

A subset $\mathbf{E}$ of the additive group of K called the set of **positive elements** such that $\mathbf{E}$ together with 0 form a semigroup, satisfying the conditions

$$\mathbf{Z}^+ \subset \mathbf{E}, \qquad \mathbf{EE} = \mathbf{E}, \qquad K = \mathbf{E} - \mathbf{E}$$

so every element of K is the difference of two elements of $\mathbf{E}$; furthermore for $e \in \mathbf{E}$ we have $\varepsilon(e) > 0$, and if $\varepsilon(e) = r$ then

$$\lambda^i(e) = 0 \text{ for } i > r \qquad \text{and} \qquad \lambda^r(e) \text{ is a unit in } K.$$

We define $\mathbf{L}$ to be the subset of elements $u \in \mathbf{E}$ such that $\varepsilon(u) = 1$. Since $\lambda^1 u = u$, it follows that $\mathbf{L}$ is a subgroup of the units K^*. Elements of $\mathbf{L}$ will be called **line elements**.

An **extension** K' of a λ-ring K is a λ-ring K' containing K, with λ^i and augmentation extending that of K, and with positive elements $\mathbf{E}'$ containing $\mathbf{E}$.

We shall be concerned with a class $\mathfrak{K}$ of λ-rings satisfying, in addition to the preceding conditions, the

Splitting Property. *For any $K \in \mathfrak{K}$ and positive element e in K, there is an extension K' of K in $\mathfrak{K}$ such that e splits in K', i.e.*

$$e = u_1 + \cdots + u_m,$$

with u_i line elements in K'.

It follows by induction that any finite set of positive elements can be simultaneously split in a suitable extension. The splitting property will allow us to deduce general formulas from the simple case of line elements. For example, the property that $\lambda^i(e) = 0$ for e positive and $i > r = \varepsilon(e)$ follows from the fact that $\lambda^i(u) = 0$ for all line elements u and $i > 1$.

More generally, for $u \in \mathbf{L}$ we have directly from the assumptions

$$\lambda_t(u) = 1 + ut,$$

and hence if e is split as above, then

$$\begin{aligned}\lambda_t(e) &= \prod_{i=1}^{r}(1 + u_i t)\\ &= 1 + \sum_{i=1}^{r} s_i(u_1, \ldots, u_r)t^i,\end{aligned}$$

where s_i is the i-th symmetric function. Since the coefficients $\lambda^i(e)$ are given *a priori* as elements of the λ-ring K, we see that the value of the symmetric function $s_i(u_1, \ldots, u_r)$ is independent of the splitting of e as a sum of line elements in K'.

For example, one sees from this formula that

$$\varepsilon(\lambda^i e) = \binom{\varepsilon(e)}{i}. \tag{1.2}$$

In other words, if $\mathbf{Z}$ is given a λ-ring structure by $\lambda^i(n) = \binom{n}{i}$, then the augmentation ε is a homomorphism of λ-rings.

Formulas for $\lambda^k(x \cdot y)$ and $\lambda^k(\lambda^j(x))$ can best be expressed in terms of certain universal polynomials P_k and $P_{k,j}$ as follows. Take independent variables $U_1, \ldots, U_m$ and $V_1, \ldots, V_n$. Let X_i be the i-th elementary symmetric polynomial in $U_1, \ldots, U_m$, and Y_i the i-th elementary symmetric polynomial in $V_1, \ldots, V_n$. For $m \geqq k$, $n \geqq k$, let

$$P_k(X_1, \ldots, X_k, Y_1, \ldots, Y_k) \in \mathbf{Z}[X_1, \ldots, X_k, Y_1, \ldots, Y_k]$$

be the polynomial of weight k in the variables X_i and in the variables Y_i (where X_i and Y_i are assigned weight i), determined by the identity

$$\text{(A)} \qquad \sum_{k \geqq 0} P_k(X_1, \ldots, X_k, Y_1, \ldots, Y_k) T^k = \prod_{i,j} (1 + U_i V_j T).$$

By setting some of the variables U_i or V_j equal to zero for $i, j > k$, one sees that the P_k are independent of the choice of m, $n \geqq k$. Similarly define

$$P_{k,j}(X_1, \ldots, X_{kj}) \in \mathbf{Z}[X_1, \ldots, X_{kj}]$$

of weight kj, by the identity for $m \geqq kj$:

$$\text{(B)} \qquad \sum_{k \geqq 0} P_{k,j}(X_1, \ldots, X_{kj}) T^k = \prod_{i_1 < \cdots < i_j} (1 + U_{i_1} \cdots U_{i_j} T).$$

Now if $x = \sum_{i=1}^{m} u_i$, $y = \sum_{j=1}^{n} v_j$, with u_i, v_j line elements, then

$$\lambda_t(x \cdot y) = \prod (1 + u_i v_j t).$$

From (A) this can be written

$$\text{(1.3)} \qquad \lambda^k(x \cdot y) = P_k(\lambda^1(x), \ldots, \lambda^k(x), \lambda^1(y), \ldots, \lambda^k(y))$$

For example, if x is a line element, then

$$\lambda^k(x \cdot y) = x^k \cdot \lambda^k y, \qquad \text{or} \qquad \lambda_t(xy) = \lambda_{xt}(y).$$

Similarly, if $x = \sum_{i=1}^{m} u_i$, then $\lambda^j(x) = \sum_{i_1 < \cdots < i_j} u_{i_1} \cdots u_{i_j}$, so

$$\lambda_t(\lambda^j x) = \prod_{i_1 < \cdots < i_j} (1 + u_{i_1} \cdots u_{i_j} t).$$

By (B) this can be written

$$\text{(1.4)} \qquad \lambda^k(\lambda^j(x)) = P_{k,j}(\lambda^1(x), \ldots, \lambda^{kj}(x)).$$

The identities (1.1)–(1.4) say that our λ-rings are what Grothendieck calls **special** λ-rings ([SGA 6], Exp. 0). This may be reinterpreted as follows. Given a commutative ring A, define $A[[T]]^{+} = TA[[T]]$, and let

$$\Lambda(A) = 1 + A[[T]]^{+}$$

be the set of power series in A with constant term 1. Define an addition in $\Lambda(A)$ by the multiplication of power series; a product $\cdot$ in $\Lambda(A)$ by the formula

$$(1 + \sum a_i t^i) \cdot (1 + \sum b_j t^j) = 1 + \sum P_k(a_1, \ldots, a_k, b_1, \ldots, b_k)t^k;$$

and λ-operations by

$$\lambda^j(1 + \sum a_i t^i) = 1 + \sum P_{k,j}(a_1, \ldots, a_{kj})t^k.$$

One verifies easily that these definitions make $\Lambda(A)$ into a special λ-ring (cf. [AT] for a readable account). For any λ-ring K,

$$\lambda_t : K \to \Lambda(K)$$

is an additive homomorphism; K is special precisely when λ_t is a homomorphism of λ-rings. Note that identities (1.1)–(1.4) hold for all elements of K, not only positive elements.

Remark. An element x in a λ-ring K is said to have **λ-dimension** $= n$ if $\lambda^i(x) = 0$ for all $i > n$, and $\lambda^n(x) \neq 0$. The ring K is called **λ-finite-dimensional** if every element is a difference of two elements of finite λ-dimension. Since positive elements have finite dimension, our axioms imply that our λ-rings are all finite dimensional. Conversely, given a λ-finite-dimensional special λ-ring K, one can define **E** to be the elements of λ-finite dimension. If one assumes that all one-dimensional elements are units, then E defines a positive structure in our sense.

Let $\sum a_i t^i$ be a power series in $K[[t]]$ with $a_0 = 1$. The coefficients of the inverse series

$$\sum b_i t^i = (\sum a_i t^i)^{-1}$$

can be determined recursively from the coefficients a_i by the relation

$$\sum_{i=0}^{k} a_i b_{k-i} = 0 \qquad \text{for} \quad k > 0.$$

For $e \in \mathbf{E}$ we define the series

$$\sigma_t(e) = \lambda_{-t}(e)^{-1} = \sum_{i=0}^{\infty} \sigma^i(e)t^i.$$

Then for each i we get a map $\sigma^i: K \to K$.

If h is a homomorphism of K into some ring and $\varphi(t) \in K[[t]]$ is a power series, then we let $h(\varphi(t))$ be the power series obtained by applying h to all the coefficients of $\varphi(t)$. In particular, we have

$$\varepsilon(\sigma_t(e))\varepsilon(\lambda_{-t}(e)) = 1.$$

Lemma 1.1. *Let $\varepsilon(e) = r + 1$. Then*

$$\varepsilon(\lambda_{-t}(e)) = (1 - t)^{r+1} \quad \textit{and} \quad \varepsilon(\sigma_t(e)) = \frac{1}{(1 - t)^{r+1}}.$$

So explicitly in terms of the coefficients,

$$\varepsilon(\lambda^i(e)) = \binom{r+1}{i} \quad \textit{and} \quad \varepsilon(\sigma^j(e)) = \binom{r+j}{j}.$$

Proof. Splitting e into $\sum u_i$, with $\varepsilon(u_i) = 1$,

$$\lambda_t(e) = \prod (1 + u_i t),$$

from which the formula for $\varepsilon(\lambda_{-t}(e))$ is clear. Since σ_t is the inverse of λ_{-t}, the formula for $\varepsilon(\sigma_t(e))$ follows. The last formula follows from the identity

$$\frac{1}{(1 - t)^{r+1}} = \sum \binom{r+j}{j} t^j.$$

I §2. An Elementary Extension of λ-Rings

Given a λ-ring K and a positive element e in K we construct a ring extension K_e of K as follows. Set $\varepsilon(e) = r + 1$,

$$p_e(T) = \sum_{i=0}^{r+1} (-1)^i \lambda^i(e) T^{r+1-i},$$

and let

$$K_e = K[T]/(p_e(T)) = K[\ell],$$

where ℓ is the image of $T \bmod p_e(T)$; we call ℓ the **canonical generator**. We have the defining relation

$$\sum_{i=0}^{r+1}(-1)^i\lambda^i(e)\ell^{r+1-i} = 0.$$

In particular, for $k \geqq r + 1$, multiplying by powers of ℓ and using $\lambda^m(e) = 0$ if $m > r + 1$, we get the relations

$$\sum_{i=0}^{k}(-1)^i\lambda^i(e)\ell^{k-i} = 0.$$

These relations translate into the single power series relation

$$\left(\sum(-1)^i\lambda^i(e)t^i\right)\left(\sum \ell^j t^j\right) = \sum_{k=0}^{r}\left(\sum_{i=0}^{k}(-1)^i\lambda^i(e)\ell^{k-i}\right)t^k.$$

Theorem 2.1. *There is a unique λ-ring structure on K_e, extending that on K, and satisfying*

$$\lambda_t(\ell) = 1 + \ell t.$$

Proof. First define a λ-ring structure on the polynomial ring $K[T]$, such that $\varepsilon(T) = 1$ and $\lambda^1(T) = T$, $\lambda^i(T) = 0$ for $i > 1$. From the fact that K is a special λ-ring it follows readily that $K[T]$ is also a special λ-ring. To show that this determines a λ-ring structure on K_e, it must be verified that the ideal $I = (p_e(T))$ is preserved by the λ-operations. Set $j = r + 1$. Then

$$(-1)^j p_e(T) = \lambda^j(e - T).$$

Using the identity (1.3) for products, one sees that it suffices to verify that

$$\lambda^k\lambda^j(e - T) \in I = (\lambda^j(e - T))$$

for all $k \geqq 1$. From the identity

$$\lambda_t(e - T) = (1 + \lambda^1(e)t + \cdots + \lambda^j(e)t^j)\cdot(1 + Tt)^{-1}$$

it follows that

$$\lambda^k(e - T) = \pm T^{k-j}\lambda^j(e - T) \in I$$

for all $k \geqq j$. Since $\lambda^k(\lambda^j(x)) = P_{k,j}(\lambda^1(x), \ldots, \lambda^{kj}(x))$, it suffices to verify that each monomial appearing in the polynomial $P_{k,j}(X_1,\ldots,X_{kj})$ contains some X_i with $i \geqq j$. To see this, simply note that

$$P_{k,j}(X_1,\ldots,X_{j-1},0,\ldots,0)$$

is identically zero, as follows from the definition (B) of $P_{k,j}$ in §1.

As seen from the proof, K_e is a special λ-ring. One may define a positive structure $\mathbf{E}_e$ on K_e generated by $\mathbf{E}$, ℓ, and $e - \ell$, i.e.

$$\mathbf{E}_e = \{\textstyle\sum a_{ij}\ell^i(e-\ell)^j \mid i,j \geqq 0, a_{ij} \in \mathbf{E}\}.$$

The elements of $\mathbf{E}_e$ with augmentation 1 are of the form $a\ell^i(e-\ell)^j$, with a a unit in $\mathbf{E}$ and $j = 0$ if $r > 1$. The equation $p_e(\ell) = 0$ shows that ℓ is a unit for all r, and that $e - \ell$ is a unit if $r = 1$. Thus $\mathbf{E}_e$ defines a positive structure on K_e, called the **canonical positive structure**.

Theorem 2.1 may be used to construct an extension K' of K in which e splits. Let $K^{(1)} = K_e$, $u_1 = \ell$, $e_1 = e - \ell$. Let $K^{(2)} = K^{(1)}_{e_1} = K^{(1)}[\ell_1]$, and set $u_2 = \ell_1$, $e_2 = e_1 - \ell_1$, and so on inductively. Then $K' = K^{(r)}$, with $e = u_1 + \cdots + u_{r+1}$.

Proposition 2.2. *Let $f_e = f\colon K_e \to K$ be the K-linear functional such that*

$$f(\ell^i) = \sigma^i(e) \qquad \textit{for} \quad 0 \leqq i \leqq r.$$

Then:

$$f(\ell^i) = \sigma^i(e) \qquad \textit{for all integers} \quad i \geqq 0;$$
$$f(\ell^{-n}) = 0 \qquad \textit{for} \quad 1 \leqq n \leqq r.$$

In fact, for $n \geqq 1$ we have the general relation

$$(-1)^{nr}\lambda^{r+1}(e)^n\ell^{-n} = \sum_{j=0}^{nr}(-1)^j\lambda^j(ne)\ell^{nr-j},$$

and $(-1)^{nr}\lambda^{r+1}(e)^n f(\ell^{-n})$ is the coefficient of t^{nr} in the power series $\lambda_{-t}(e)^{n-1}$.

Proof. We apply f to the power series relation and use K-linearity to get

$$(\textstyle\sum(-1)^i\lambda^i(e)t^i)(\sum f(\ell^j)t^j) = \displaystyle\sum_{k=0}^{r}\left(\sum_{i=0}^{k}(-1)^i\lambda^i(e)\sigma^{k-i}(e)\right)t^k = 1,$$

so $f(\ell^j) = \sigma^j(e)$ for $j \geqq 0$. This gives the values of f on positive powers of ℓ.

Next we look at negative powers of ℓ. For ℓ^{-n} with $n = 1$ the stated relation is the equation for ℓ, namely

$$(-1)^r \lambda^{r+1}(e)\ell^{-1} = \sum_{i=0}^{r} (-1)^i \lambda^i(e)\ell^{r-i}.$$

The general relation follows by induction, directly from the definition of a λ-ring. We apply f to the values obtained in the first part of the proposition. Then we find that

$$(-1)^{nr}\lambda^{r+1}(e)^n f(\ell^{-n})$$

is the coefficient of t^{nr} in the power series

$$\lambda_{-t}(ne)\sigma_t(e) = \lambda_{-t}(e)^n\sigma_t(e) = \lambda_{-t}(e)^{n-1}.$$

Since $\lambda^i(e) = 0$ for $i > r + 1$, the coefficient of t^{nr} in this power series is 0 if

$$nr > (n - 1)(r + 1),$$

which gives precisely $n < r + 1$, or $n \leqq r$. This concludes the proof.

Remark. The factor $(-1)^{nr}\lambda^{r+1}(e)^n$ is a unit in K, so the second part of the proposition gives the values of f at negative powers of ℓ.

The functional

$$f_e\colon K_e \to K$$

such that $f_e(\ell^i) = \sigma^i(e)$ will be called the **functional associated with the extension** K_e of K. As we did above, if e is fixed throughout a discussion, we omit the subscript e and write simply f. Although we have no immediate use for it, we give immediately the following application.

Corollary 2.3. *Let* $q = e - \ell$. *Then*

$$f_e(\lambda_t(q)) = 1.$$

Proof. We have

$$\begin{aligned} f(\lambda_t(q)) &= f(\lambda_t(e)(\lambda_t(\ell))^{-1}) \\ &= \lambda_t(e) \sum_{i=0}^{\infty} (-1)^i f(\ell^i) t^i \\ &= \lambda_t(e) \sigma_{-t}(e) \\ &= 1, \end{aligned}$$

as was to be shown.

Remark. For a λ-finite-dimensional λ-ring K, the considerations of the preceding sections show that the following are equivalent:

(i) K is a special λ-ring.
(ii) Every e in K with finite λ-dimension splits in some λ-ring extension of K.
(iii) Every e in K with finite λ-dimension splits in some special λ-ring extension of K.
(iv) For every e in K with finite λ-dimension $K_e = K[\ell]$ is a special λ-ring, with $\lambda_t(\ell) = 1 + \ell t$.

If a class $\mathfrak{K}$ of λ-finite-dimensional λ-rings contains K_e for each $K \in \mathfrak{K}$ and each $e \in K$ of finite λ-dimension, then the splitting principle holds for all K in $\mathfrak{K}$, and all K in $\mathfrak{K}$ are special.

I §3. Chern Classes and the Splitting Principle

The formalism of symmetric functions, Chern class homomorphisms, and the splitting principle (for instance as in this section) were used and developed for the first time by Hirzebruch [Hi] for the proof of the Hirzebruch–Riemann–Roch theorem.

Let A be a graded (commutative) ring,

$$A = \bigoplus_{i=0}^{\infty} A^i \quad \text{and} \quad A^+ = \bigoplus_{i=1}^{\infty} A^i,$$

where A^i is the i-th graded component of A. Let

$$\Lambda^\circ(A) = \{1 + a_1 t + a_2 t^2 + \cdots\}$$

be the group of formal power series with leading term 1 and $a_i \in A^i$. Let K be a λ-ring, and let

$$c_t \colon K \to \Lambda^\circ(A)$$

be a homomorphism of abelian groups. We write

$$c_t(x) = \sum c^i(x)t^i$$

with $c^i(x) \in A^i$. The fact that c_t is a homomorphism can be written:

$$c^k(x + y) = \sum_{i+j=k} c^i(x)c^j(y). \tag{3.1}$$

We call c_t a **Chern class homomorphism** with values in A, if in addition, it satisfies properties **CC 1**, **CC 2**, **CC 3** below, and the splitting principle which follows.

We require the following conditions for line elements L:

CC 1. For $u \in \mathbf{L}$, $c^i(u) = 0$ for $i > 1$, that is

$$c_t(u) = 1 + c^1(u)t.$$

CC 2. For $u, v \in \mathbf{L}$ we have

$$c^1(uv) = c^1(u) + c^1(v).$$

In other words, $c^1 \colon \mathbf{L} \to A^1$ is a homomorphism.

The i-th graded component $c^i(x)$ is called the i-th **Chern class**. The third condition mentioned above is then:

CC 3. For all $e \in \mathbf{E}$ and all $i \geqq 1$, $c^i(e)$ is nilpotent.

Remark. The formal theory of Chern classes actually does not require the nilpotence until Chapter III, §3 and §4, where an even stronger condition will be imposed. If we do not require nilpotence, then instead of the graded ring A we should take the ring

$$\hat{A} = \prod_{i=0}^{\infty} A^i$$

which is the completion of A, and consists of all formal series $\sum a_i$, with $a_i \in A^i$.

Under **CC 3**, given $x \in K$ it will follow that all $c^i(x)$ vanish for large i, and then

$$c_t(x) = \sum c^i(x)t^i$$

is called the **Chern polynomial** of x. The class

$$c(x) = \sum c^i(x)$$

in A is called the **total Chern class** of x. If we did not assume **CC 3**, then $c(x)$ would lie in $\hat{A}$. We then have a homomorphism

$$c: K \to 1 + A^+, \qquad c(x) = 1 + \sum c^i(x).$$

We often write simply

$$c: K \to A$$

for this homomorphism from the additive group of K to the multiplicative group of units of A. The variable t is convenient in order to keep track of the grading. Furthermore, representing the Chern class homomorphism as a polynomial c_t exhibits better the formal analogy with λ-rings and the power series λ_t. Treating t as a variable also allows us to substitute special values, like $t = -1$, so that for instance we get

$$c_{-1}(u) = 1 - c^1(u).$$

Both notations, with and without the variable t, are useful for applications.

We shall also require a

Splitting Principle. *Given a finite set $\{e_i\}$ of positive elements of K, there is a λ-ring extension K' of K in which each e_i splits, such that c extends to a homomorphism*

$$c: K' \to A'$$

for some graded extension A' of A.

Let us note some consequences of this splitting principle. If e splits into

$$e = u_1 + \cdots + u_m,$$

let $a_i = c^1(u_i)$. Then we get a factorization

$$c_t(e) = \prod_{i=1}^{m} (1 + a_i t)$$

of the Chern polynomial into linear factors. In particular

$$c^i(e) = 0 \qquad \text{for} \quad i > m.$$

In addition,

$$c^k(e) = s_k(a_1, \ldots, a_m) = s_k(c^1(u_1), \ldots, c^1(u_m))$$

is the k-th elementary symmetric function of the first Chern classes of $u_1, \ldots, u_m$. The a_i are called **Chern roots** for e. The equation shows that any symmetric polynomial in the Chern roots can be written as a polynomial in the Chern classes of e, and that the resulting expression is independent of the splitting of e.

If also $f = \sum_{j=1}^{n} v_j$, with $b_j = c^1(v_j)$ Chern roots for f, then

$$c_t(e \cdot f) = \prod_{i,j} (1 + (a_i + b_j)t). \tag{3.2}$$

In other words, the Chern roots for a product are pairwise sums of Chern roots for the factors. In particular, if $f = v$ is a line element, one has the useful explicit formula

$$c_t(e \cdot v) = \sum_{i=0}^{m} t^i c_t(v)^{m-i} c^i(e),$$

or

$$c^k(e \cdot v) = \sum_{j=1}^{k} \binom{m-j}{k-j} c^j(e)(c^1(v))^{k-j}.$$

If $m = \varepsilon(e)$ we call

$$c^{\text{top}}(e) = c^m(e) = \prod_{i=1}^{m} a_i$$

the **top Chern class** of e; it is often convenient to omit the value $\varepsilon(e) = m$ from the notation.

There are similar formulas for Chern classes of $\lambda^j e$, since

$$c_t(\lambda^j e) = \prod_{i_1 < \cdots < i_j} (1 + (a_{i_1} + \cdots + a_{i_j})t). \tag{3.3}$$

As in §1, these formulas may be expressed in terms of universal polynomials. Note, however, that the formulas for $c^k(e \cdot f)$ and $c^k(\lambda^j e)$ depend on the ranks $\varepsilon(e)$ and $\varepsilon(f)$ as well as on their Chern classes. To avoid this complication, we shall consider the restriction of c_t to $\tilde{K} = \mathrm{Ker}(\varepsilon)$, the elements of augmentation zero.

In §1 we used universal polynomials to construct a λ-ring $\Lambda(A)$. One sees from the definition that the product and λ^i take $\Lambda^\circ(A)$ into itself, so $\Lambda^\circ(A)$ becomes a λ-ring, but without unit. The identities (3.1)–(3.3) imply that

$$c_t\colon \tilde{K} \to \Lambda^\circ(A)$$

is a homomorphism of λ-rings without unit.* We say that c_t is a **λ-homomorphism** in this case.

Formulas (3.2) and (3.3) were deduced from the splitting principle. We shall see that, conversely, these identities imply the splitting principle, in an explicit form that will be useful later.

We shall now construct a graded ring extension of A in a way similar to the construction of K_e from K in §2. First note that the polynomial ring $A[W]$ (where W is a variable) has a unique grading extending that of A such that W has degree 1.

Given an element $c = 1 + \sum c_i t^i \in \Lambda^\circ(A)$ and an integer m such that $c_i = 0$ for $i > m$, let $p_c(W)$ be the polynomial

$$p_c(W) = W^m - c_1 W^{m-1} + \cdots \pm c_m = \sum_{i=0}^{m} (-1)^i c_i W^{m-i}.$$

Then $p_c(W)$ is homogeneous, and the factor ring

$$A_c = A[W]/(p_c(W)) = A[w],$$

defines a graded ring extension of A. The element $w = W \bmod (p_c(W))$ is called the **canonical generator** of A_c.

For later use, we define the **associated functional**

$$g_c\colon A_c \to A$$

to be the A-linear homomorphism such that

$$g_c(w^j) = \begin{cases} 0 & \text{for } 0 \leqq j < m - 1, \\ 1 & \text{for } j = m - 1. \end{cases}$$

* One can also construct a λ-ring structure on $\mathbf{Z} \times \Lambda^\circ(A)$, so that

$$K \to \mathbf{Z} \times \Lambda^\circ(A), \qquad x \mapsto (\varepsilon(x), c_t(x))$$

is a homomorphism of λ-rings with unit (cf. [SGA 6], p. 30 ff.).

Theorem 3.1. *Suppose $c_t\colon K \to \Lambda^\circ A$ is a λ-homomorphism, and e is a positive element in K such that $\varepsilon(e) = m$ and $c^i(e) = 0$ for $i > m$. Then c_t extends uniquely to a λ-homomorphism*

$$c_t\colon K_e \to \Lambda^\circ(A_{c(e)}),$$

such that if ℓ and w are the canonical generators for K_e and $A_{c(e)}$, respectively, then

$$c_t(\ell) = 1 + wt.$$

Proof. Let $K[T]$ be the λ-ring extension defined in the proof of Theorem 2.1, and extend c_t to a homomorphism

$$c_t\colon K[T] \to \Lambda^\circ(A[W])$$

by setting $c_t(T) = 1 + WT$. It is straightforward to verify that this extension is also a λ-homomorphism.

To conclude the proof, we must show that for $k \geqq 1$, $c^k(p_e(T)) \in J$, where J is the ideal in $A[W]$ generated by $p_{c(e)}(W)$. Equivalently, we must show that

$$c^k(\lambda^m(e - T)) \in J.$$

Since c_t is a λ-homomorphism, we get

$$c_t(\lambda^m(e - T)) = \lambda^m(c_t(e - T)) = \lambda^m(a),$$

where $a = c_t(e)/(1 + Wt)$. Therefore $a = 1 + \sum a_k t^k$, with

$$a_k = (-W)^{k-m}(c^m(e) - c^{m-1}(e)W + \cdots + (-1)^m W^m)$$

for all $k \geqq m$. Therefore $a_k \in J$ for all $k \geqq m$. Since

$$\lambda^m(a) = 1 + \sum P_{k,m}(a_1, \ldots, a_{km})t^k,$$

and, as we saw in the proof of Theorem 2.1, each monomial in $P_{k,m}(a)$ is divisible by some a_i for $i \geqq m$, it follows that $P_{k,m}(a) \in J$, as required.

I §4. Chern Character and Todd Classes

The splitting principle allows us to go further, by using systematically the factorization of the Chern polynomial $c_t(e)$ in linear factors $1 + a_i t$. Let

$$\varphi(t) \in \mathbf{Z}[[t]] \qquad \text{or} \qquad \varphi(t) \in \mathbf{Q}[[t]]$$

be a power series with integer coefficients, or rational coefficients if A is also a $\mathbf{Q}$-algebra. To each such power series we can associate an additive homomorphism

$$\mathrm{ch}_\varphi : K \to A$$

as follows. We first define ch_φ on $\mathbf{E}$, by setting

$$\mathrm{ch}_\varphi(e) = \sum_{i=1}^{m} \varphi(a_i).$$

Since we assume that the first Chern classes a_i are nilpotent, the evaluation of the power series $\varphi(a_i)$ is defined, and is a polynomial in a_i for each i. Furthermore, the value on the right-hand side is independent of the choice of the splitting. To see this, note that if $W_1, \ldots, W_m$ are new independent variables, then

$$\sum_{i=1}^{m} \varphi(W_i t) = \sum_{i=0}^{\infty} H_j(s_1, \ldots, s_j) t^j,$$

where H_j is a polynomial of weight j with rational coefficients, and s_j is the j-th elementary function of $W_1, \ldots, W_m$. We call H_j the **associated Hirzebruch polynomials**. Then

$$\mathrm{ch}_\varphi(e) = \sum H_j(c^1(e), \ldots, c^j(e)).$$

It follows immediately that for $e, e' \in \mathbf{E}$ we have

$$\mathrm{ch}_\varphi(e + e') = \mathrm{ch}_\varphi(e) + \mathrm{ch}_\varphi(e'),$$

so ch_φ is a homomorphism on the semigroup of elements of $\mathbf{E}$. For any element $x = e - e'$ of K we define

$$\mathrm{ch}_\varphi(x) = \mathrm{ch}_\varphi(e) - \mathrm{ch}_\varphi(e').$$

It is trivially verified that this is well defined, i.e. independent of the representation of x as a difference of elements in $\mathbf{E}$, and that ch_φ is a homomorphism of K into A. Explicitly,

$$\mathrm{ch}_\varphi(x) = \sum H_j(c^1(x), \ldots, c^j(x)).$$

The most important example is the **Chern character** written without subscript

$$\mathrm{ch} : K \to A$$

such that $\mathrm{ch} = \mathrm{ch}_\varphi$, where φ is the exponential power series

$$\varphi(t) = \exp(t) = \sum \frac{t^k}{k!}.$$

In this case, A must be a **Q**-algebra, or we tensor A with **Q**. Then by definition, if $e = \sum_{i=1}^{m} u_i$ and $a_i = c^1(u_i)$, we have

$$\mathrm{ch}(e) = \sum_{i=1}^{m} \sum_{k=0}^{\infty} \frac{a_i^k}{k!}.$$

Proposition 4.1. *The Chern character* $\mathrm{ch}\colon K \to A$ *is a ring homomorphism.*

Proof. It suffices to verify this for products of elements in **E**. Say $e = \sum u_i$ and $e' = \sum v_j$. Then $ee' = \sum u_i v_j$, and

$$\begin{aligned} \mathrm{ch}(ee') &= \sum_{i,j} \exp(c^1(u_i v_j)) \\ &= \sum_{i,j} \exp(c^1(u_i) + c^1(v_j)) \\ &= \mathrm{ch}(e)\,\mathrm{ch}(e') \end{aligned}$$

as desired.

Of course, we also have $\mathrm{ch}(1) = 1$, as follows directly from the definition of $c^1(1) = 0$.

For the Chern character ch, the first few Hirzebruch polynomials H_j as mentioned above can be calculated to be:

$$\begin{aligned} H_1(s_1) &= s_1, \\ H_2(s_1, s_2) &= \tfrac{1}{2}(s_1^2 - 2s_2), \\ H_3(s_1, s_2, s_3) &= \frac{1}{3!}(s_1^3 - 3s_1 s_2 + 3s_3), \\ H_4(s_1, s_2, s_3, s_4) &= \frac{1}{4!}(s_1^4 - 4s_1^2 s_2 + 4s_1 s_3 + 2s_2^2 - 4s_4). \end{aligned}$$

Remark on notation. The Chern classes are usually denoted by c_i instead of c^i. To lay down the general formalism we thought it better to preserve the upper numbering, in order not to break the notational analogy, say of the power series c_t with λ_t. However, we now see that this upper numbering is extremely disagreeable if we wish to substitute the

Chern classes in the Hirzebruch polynomials, because we have to take ordinary powers. Hence in practice, one may revert to the lower numbering and write for instance

$$\mathrm{ch}(e) = \varepsilon(e) + c_1 + \tfrac{1}{2}(c_1^2 - 2c_2) + \dots.$$

We can perform a similar construction multiplicatively. Let

$$\varphi(t) \in 1 + t\mathbf{Z}[[t]] \qquad \text{or} \qquad \varphi(t) \in 1 + t\mathbf{Q}[[t]]$$

be a power series with constant term 1 and integer coefficients, or rational coefficients if A is a **Q**-algebra. Then we define the corresponding **Todd homomorphism** on positive elements by

$$\mathrm{td}_\varphi(e) = \prod_{i=1}^{m} \varphi(a_i).$$

If $W_1, \dots, W_m$ are independent variables, then we can write

$$\prod_{i=1}^{m} \varphi(W_i t) = \sum_{j=0}^{\infty} Q_j(s_1, \dots, s_j)t^j,$$

where $Q_j(s_1, \dots, s_j)$ is a polynomial of weight j with integer (resp. rational) coefficients in the elementary symmetric functions $s_1, \dots, s_m$ of $W_1, \dots, W_m$. Again we call Q_j the **associated Hirzebruch polynomials**. Then

$$\mathrm{td}_\varphi(e) = \sum_{j=0}^{\infty} Q_j(c^1(e), \dots, c^j(e))$$

is independent of the splitting of e. Thus

$$\mathrm{td}_\varphi : K \to 1 + A^+$$

is a homomorphism from the additive group of K into the multiplicative group of units of A, and in fact those units which are of the form $1 + b$ with b nilpotent.

If $\varphi = \beta$ is the power series

$$\beta(t) = \frac{t\mathbf{e}^t}{\mathbf{e}^t - 1}, \qquad \text{where} \quad \mathbf{e}^t = \exp(t).$$

then we write $\mathrm{td}(e)$ instead of $\mathrm{td}_\beta(e)$, and call this simply "**the**" **Todd homomorphism**, determined by the original data of a Chern class homomorphism. In this case, A must be a $\mathbf{Q}$-algebra, or we tensor A with $\mathbf{Q}$. The first few Hirzebruch polynomials Q_j can be calculated to be:

$$\begin{aligned}
Q_1(s_1) &= \tfrac{1}{2}s_1,\\
Q_2(s_1, s_2) &= \tfrac{1}{12}(s_1^2 + s_2),\\
Q_3(s_1, s_2, s_3) &= \tfrac{1}{24}s_1 s_2,\\
Q_4(s_1, s_2, s_3, s_4) &= -\tfrac{1}{720}(s_1^4 - 4s_1^2 s_2 - 3s_2^2 - s_1 s_3 + s_4).
\end{aligned}$$

Generalizations. 1. Even without the assumption that the $c^i(x)$ are nilpotent, one can define a homomorphism

$$\mathrm{ch}_{\varphi,t} K \to A[[t]]$$

by $\mathrm{ch}_{\varphi,t}(x) = \sum H_j(c^1(x), \ldots, c^j(x))t^i$. For the ordinary Chern character,

$$\mathrm{ch}_t\colon K \to A[[t]]$$

is a ring homomorphism, as is

$$\mathrm{ch}\colon K \to \hat{A},$$

where $\hat{A}$ is the completion of A.

2. For Todd classes of positive elements e it is not necessary to assume that the constant term of φ is 1. One may define

$$\mathrm{td}_{\varphi,t}(e) = \varphi(0)^{\varepsilon(e)} + \sum_{j\geqq 1} Q_j(c^1(e), \ldots, c^j(e))t^j.$$

This $\mathrm{td}_{\varphi,t}$ will take sums of positive elements to products. If φ is a polynomial, or if the $c^i(e)$ are nilpotent, then $\mathrm{td}_\varphi(e) \in A$. If $\varphi(0)$ is a unit in A, then td_φ extends to a homomorphism on all of K.

3. With a systematic use of symmetric functions and Hirzebruch polynomials, one may avoid any explicit use of a splitting principle.

I §5. Involutions

We shall be concerned with λ-rings K which have an **involution**, by which we mean a homomorphism $x \mapsto x^\vee$ from the ring K to itself, satisfying

$$x^{\vee\vee} = x, \qquad \varepsilon(x^\vee) = \varepsilon(x), \qquad \text{and} \qquad u^\vee = u^{-1} \text{ for } u \in \mathbf{L}.$$

We assume also that any positive element can be split in some extension K' to which the involution extends.

Lemma 5.1. *Let $e \in \mathbf{E}$ and $\varepsilon(e) = m$. Then for all i with $0 \leqq i \leqq m$ we have*

$$\lambda^i(e) = \lambda^{m-i}(e^\vee)\lambda^m(e).$$

Proof. By definition, using a splitting, we get

$$\begin{aligned}
\lambda_t(e) = \sum \lambda^i(e)t^i &= \prod_{i=1}^{m} (1 + u_i t) \\
&= \prod_{i=1}^{m} (u_i t)(u_i^{-1} t^{-1} + 1) \\
&= \prod_{i=1}^{m} u_i \cdot t^m \prod_{i=1}^{m} (1 + u_i^\vee t^{-1}) \\
&= \prod_{i=1}^{m} u_i \cdot t^m \sum_{i=0} \lambda^i(e^\vee)t^{-i} \\
&= \lambda^m(e) \sum_{i=0}^{m} \lambda^i(e^\vee)t^{m-i},
\end{aligned}$$

which concludes the proof.

Conversely, if the formula of Lemma 5.1 is valid for e, then the involution $\vee$ extends to an involution of K_e, with $\ell^\vee = \ell^{-1}$. This follows from the equation

$$\sum (-1)^{m-i}\lambda^{m-i}(e^\vee)(\ell^{-1})^i = (-1)^m\lambda^m(e)^\vee \ell^{-m}\left(\sum (-1)^i\lambda^i(e)\ell^{m-i}\right).$$

If $c_t \colon K \to \Lambda^\circ(A)$ is a Chern class homomorphism, then

$$c^1(u^\vee) = c^1(u^{-1}) = -c^1(u)$$

for a line element. From the splitting principle it follows that

$$c_i(x^\vee) = (-1)^i c_i(x) \tag{5.1}$$

for all $x \in K$.

It follows that

$$\mathrm{ch}(x^\vee) = -\mathrm{ch}(x). \tag{5.2}$$

Another simple formula which follows easily from the splitting principle is

Proposition 5.2. *For a positive element e,*

$$\mathrm{td}(e^\vee) = \mathrm{td}(e)\exp(-c^1(e)).$$

Our main interest, however, lies in the next formula, which embodies a Riemann–Roch relation as will be seen in Chapter II, Theorem 2.1.

Proposition 5.3. *For a positive element e we have*

$$\mathrm{td}(e)\,\mathrm{ch}(\lambda_{-1}(e^\vee)) = c^{\mathrm{top}}(e),$$

where $\mathrm{ch}(\lambda_{-1}(e^\vee)) = \sum(-1)^i\,\mathrm{ch}\,\lambda^i(e^\vee)$. *Or in other words,*

$$\mathrm{ch}(\lambda_{-1}(e^\vee)) = c^{\mathrm{top}}(e)\,\mathrm{td}(e)^{-1}.$$

Proof. By definition, splitting $e = \sum_{i=j}^{m} u_i$, with $a_i = c^1(u_i)$, we have

$$\mathrm{td}(e) = \prod_i \beta(a_i) = \prod_i \frac{a_i\mathbf{e}^{a_i}}{\mathbf{e}^{a_i}-1}.$$

Also,

$$\lambda_t(e^\vee) = \prod_i (1 + u_i^\vee t),$$

whence

$$\mathrm{ch}\,\lambda_t(e^\vee) = \prod_i (1 + \mathrm{ch}(u_i^\vee)t)$$
$$= \prod_i (1 + \mathbf{e}^{-a_i}t)$$

and therefore

$$\mathrm{ch}\,\lambda_{-1}(e^\vee) = \prod_i (1 - \mathbf{e}^{-a_i})$$
$$= \prod_i (\mathbf{e}^{a_i} - 1)/\mathbf{e}^{a_i}.$$

Multiplying, we get

$$\mathrm{td}(e)\,\mathrm{ch}(\lambda_{-1}(e^\vee)) = \prod_{i=1}^{m} a_i = c^m(e).$$

This proves the proposition.

I §6. Adams Operations

We return to a single λ-ring K. We define the **Adams power series** and the **Adams operations** $\psi^j: K \to K$ by the formula

$$\psi_t(x) = \varepsilon(x) - t\frac{d}{dt}\log \lambda_{-t}(x) = \sum_{j=0}^{\infty} \psi^j(x)t^j.$$

Proposition 6.1.

(i) *If* $u \in \mathbf{L}$, *then* $\psi^j(u) = u^j$ *for all* j.
(ii) *For all* j, *the map* ψ^j *is a ring homomorphism.*
(iii) $\psi^i(\psi^j(x)) = \psi^{ij}(x)$ *for all* $x \in K$ *and all* i, j.

Proof. The first assertion is immediate. For the second, it suffices to prove the homomorphic property for elements of $\mathbf{E}$. The fact that ψ^j is additive is immediate, and that it is also a multiplicative homomorphism follows by splitting an element of $\mathbf{E}$ as usual, and by using the first assertion. The third statement is then clear since the desired relation is true on elements $x = u$ in $\mathbf{L}$. This concludes the proof.

Since ψ^j is a ring homomorphism like the Chern character, we may call it an **Adams character** rather than Adams operation. We can also write:

$$\frac{d}{dt}\log \lambda_t(x) = \sum_{j=1}^{\infty} (-1)^{j-1}\psi^j(x)t^{j-1}.$$

If $e \in \mathbf{E}$ is a positive element, and $e = \sum_{i=1}^{m} u_i$ is a splitting, then

$$\lambda_t(e) = \prod (1 + u_i t)$$

so

$$\frac{d}{dt}\log \lambda_t(e) = \sum \frac{u_i}{1 + u_i t} = \sum (-1)^{j-1}(u_1^j + \cdots + u_m^j)t^{j-1}.$$

Therefore, if N_j is the **(Hirzebruch–Newton) polynomial** with integer coefficients such that

$$W_1^j + \cdots + W_m^j = N_j(s_1, \ldots, s_m),$$

where $s_1, \ldots, s_m$ are the elementary symmetric functions of $W_1, \ldots, W_m$, then

$$\psi^j(e) = N_j(\lambda^1(e), \ldots, \lambda^m(e)).$$

Let $\varphi(t)$ be a polynomial, say with integer coefficients, and constant term equal to 1. In the present context there is a **Todd homomorphism**

$$\mathrm{td}_\varphi : \mathbf{E} \to K$$

from the additive monoid of positive elements to the multiplicative monoid of elements of K, by the same method as before. From a splitting of e we let

$$\mathrm{td}_\varphi(e) = \prod \varphi(u_i).$$

The value is independent of the splitting, and is a universal polynomial in $\lambda^1(e), \ldots, \lambda^r(e)$, determined by φ alone. If $\varphi(u)$ is a unit for each line element u, then td_φ extends to a homomorphism from the additive group K to the multiplicative group K^* (see Generalization 2 of §4).

Let j be an integer $\geqq 1$. We let $\theta^j = \mathrm{td}_{\varphi_j}$ where $\varphi_j(t)$ is the polynomial

$$\varphi_j(t) = 1 + t + \cdots + t^{j-1} = \frac{1 - t^j}{1 - t}.$$

Thus by definition,

$$\theta^j(e) = \prod_{i=1}^{m} (1 + u_i + \cdots + u_i^{j-1}).$$

The classes $\theta^j(e)$ are known as "**Bott's cannibalistic classes**". If it happens that j is a unit in K, then $\theta^j(e)$ is a unit, and θ^j extends to all of K. The next result is an analogue of Proposition 5.3, and will be interpreted as a Riemann–Roch theorem in Chapter II, Theorem 3.1.

Proposition 6.2. *For a positive element e we have*

$$\psi_j(\lambda_{-1}(e)) = \lambda_{-1}(e)\theta^j(e).$$

Proof. Using the splitting, we get

$$\begin{aligned}\psi^j(\lambda_{-1}(e)) &= \psi_j\left(\prod_i (1 - u_i)\right) = \prod_i (1 - u_i^j)\\ &= \prod_i (1 - u_i) \prod_i (1 + u_i + \cdots + u_i^{j-1})\\ &= \lambda_{-1}(e)\theta^j(e),\end{aligned}$$

as was to be shown.

The following proposition also follows immediately from the definitions.

Proposition 6.3. *Let $c\colon K \to A$ be a Chern class homomorphism. Then for all integers $j \geqq 1$ and $k \geqq 1$ we have*

$$\mathrm{ch}^k\, \psi^j(x) = j^k\, \mathrm{ch}^k(x),$$

where ch^k *is the* k-th *graded component of* ch.

On may also define ψ^j for $j < 0$ by the formula

$$\psi^j(x) = \psi^{-j}(x^\vee),$$

so that Proposition 6.1 continues to hold for this extended family of operations. We shall not need ψ^j for negative j, however.

For a discussion of Adams operations on representation rings of finite groups, see [Ke] and [Kr].

CHAPTER II

Riemann–Roch Formalism

This chapter deals with the axiomatization of the functorial properties of the Grothendieck group $K(X)$.

The covariant and contravariant functorial properties of the K-functor, and another related graded ring functor $A(X)$, are such that to prove the Riemann–Roch formula it suffices to do so for morphisms which generate the category. In geometry, there are two types of morphisms to which one reduces the proof:

regular imbeddings;

projections from a projective bundle $\mathbf{P}(E)$.

In Chapter IV we describe the geometry of these morphisms. The regular imbeddings are local complete intersections. Among these are the elementary imbeddings which are the zero sections of a vector bundle. It turns out that any regular imbedding has a deformation to an elementary imbedding into the normal bundle. In Chapter V, we derive basic functorial properties of such morphisms on the K-group. They have simple algebraic formulations, and it turns out that these simple algebraic properties suffice to give a proof of the Riemann–Roch formula. For example, in Chapter V, Proposition 4.3, we show that for a regular section f of a vector bundle E, if we let $e = [E]$ be its class in the K-group, then

$$f_K(1) = \lambda_{-1}(e^\vee),$$

where $e^\vee$ is the class of the dual bundle. We take this, and the analogous formula on the graded ring functor A, as the abstract definition of an elementary imbedding in the present Chapter II, §2. The essential part of the proof of Riemann–Roch for such a morphism, depending only on this property, was given in Proposition 5.3 of Chapter I.

Similarly, Chapter V, Theorem 2.3 and Corollary 2.4 give the basic structure of the K-algebra for a projective bundle. This structure was axiomatized in Chapter I, §2, and the Riemann–Roch formula using only these axioms is then proved in the present Chapter II, Theorem 2.2.

Therefore readers may profitably read simultaneously Chapter V and Chapters I and II.

For a projective variety X, the ring $A(X)$ can be taken to be the Chow ring of cycles modulo rational equivalence, tensored with $\mathbf{Q}$. This requires more algebraic geometry, for which we refer to [F 2]. In [SGA 6], Grothendieck showed how one could define a filtration in $K(X)$ and how the associated graded algebra (tensored with $\mathbf{Q}$) could be used instead of the Chow ring. We have taken this graded ring for $A(X)$ for the main statement of the Grothendieck Riemann–Roch theorem given in Chapter V, Theorem 4.3, complemented by the more geometric comments of Chapter VI, §5, especially Propositions 5.4 and 5.5 which relate the Grothendieck filtration to filtration by codimension. However, the axiomatization of Chapter II, §1 and §2, provides the algebraic formalism for other situations. Again, readers should compare immediately these two parts of the book, and the discussions of Chapter VI (giving other geometric contexts) to get a better feeling both for the underlying algebra, and the geometric applications which motivated it.

Despite the fact that the algebraic formalism of the first three chapters originated in the theory of vector bundles, it exists independently of that theory, and is applicable to the theory of group representations. An algebraist who wishes to disregard topology or vector bundles may therefore still understand the first three chapters without having to go through the algebraic geometry of Chapters IV and V. The fundamental reason why the general algebra was placed first was to exhibit clearly its independence from any of the multiple contexts in which it may be applied. For the context of group representations, we refer the reader to various papers of the Bibliography by Atiyah–Hirzebruch, Evens–Kahn, Grothendieck, Knopfmacher, Thomas.

II §1. Riemann–Roch Functors

It is now convenient to view the objects we have defined so far in a functorial setting. We start with a category $\mathfrak{C}$. We shall be concerned with functors on $\mathfrak{C}$ which are simultaneously contravariant and covariant. Such a functor H assigns to each object X in $\mathfrak{C}$ a ring $H(X)$, and to each morphism $f: X \to Y$ in $\mathfrak{C}$ homomorphisms*

$$f^H: H(Y) \to H(X) \qquad \text{and} \qquad f_H: H(X) \to H(Y)$$

* Homomorphisms like f^H and f_H are usually denoted f^* and f_*. The more explicit notation is useful for Riemann–Roch, where several such functors are considered simultaneously.

satisfying the following conditions:

F 1. *$X \mapsto H(X)$ is a contravariant functor fom $\mathfrak{C}$ to rings via f^H.*

F 2. *$X \mapsto H(X)$ is a covariant functor from $\mathfrak{C}$ to abelian groups via f_H.*

F 3. *The* **projection formula** *holds, that is for all morphisms $f: X \to Y$, and all $x \in H(X)$, $y \in H(Y)$ we have*

$$f_H(x \cdot f^H(y)) = f_H(x) \cdot y.$$

An important special case of the projection formula is the formula

$$f_H(f^H(y)) = f_H(1)y. \tag{1.1}$$

By a **Riemann-Roch functor** we mean a triple (K, ρ, A), where K and A are functors satisfying **F 1** to **F 3**, and

$$\rho: K \to A$$

is a morphism of contravariant functors, i.e. for each X, $\rho_X: K(X) \to A(X)$ is a ring homomorphism, and

$$f^A \rho_Y(y) = \rho_X(f^K(y))$$

for all $f: X \to Y$, $y \in K(Y)$.

We shall call ρ the **Riemann-Roch character**. In special cases it may bear other names such as Chern character or Adams character, to emphasize the special features as they arise. These special cases will be dealt with in subsequent sections.

We shall say that **Riemann-Roch holds** for a morphism f if, for some element $\tau_f \in A(X)$,

$$\rho_Y f_K(x) = f_A(\tau_f \cdot \rho_X(x))$$

for all $x \in K(X)$. That is, the diagram

$$\begin{array}{ccc} K(X) & \xrightarrow{\tau_f \cdot \rho} & A(X) \\ {\scriptstyle f_K}\downarrow & & \downarrow{\scriptstyle f_A} \\ K(Y) & \xrightarrow{\rho} & A(Y) \end{array}$$

is commutative. As we have done, it is customary to omit the subscripts, writing ρ in place of ρ_X or ρ_Y.

The factor τ_f measures the extent to which ρ fails to be covariantly functorial. We call τ_f the **Riemann-Roch multiplier**, or simply the **multiplier**. When precision is necessary we say that **Riemann-Roch holds for** f

with respect to (K, ρ, A) **with multiplier** τ_f, if the preceding diagram is commutative.

Next we give some general criteria for Riemann-Roch to hold.

Theorem 1.1. *Let* $f: X \to Y$ *and* $g: Y \to Z$ *be morphisms. Assume that Riemann-Roch holds for* f *and* g *with multipliers* τ_f *and* τ_g. *Then Riemann-Roch holds for* $g \circ f$ *with multiplier*

$$\tau_{g \circ f} = f^A(\tau_g) \cdot \tau_f.$$

Proof. The routine is as follows:

$$\begin{aligned}
\rho_Z(g_K f_K(x)) &= g_A(\tau_g \cdot \rho_Y f_K(x)) && \text{by R-R for } g \\
&= g_A(\tau_g \cdot f_A(\tau_f \cdot \rho_X(x))) && \text{by R-R for } f \\
&= g_A f_A(f^A(\tau_g) \cdot \tau_f \cdot \rho_X(x)) && \text{by projection formula,}
\end{aligned}$$

thus proving the theorem.

The next criterion will apply to certain types of imbeddings, first in the abstract context of Chapter II, Theorem 2.1, and then to geometric situations like Chapter V, Proposition 4.3.

Theorem 1.2. *If* $f^K: K(Y) \to K(X)$ *is surjective, and there is an element* τ *in* $A(Y)$ *such that*

$$\rho_Y(f_K(1)) = f_A(1)\tau,$$

then Riemann-Roch holds for f *with multiplier*

$$\tau_f = f^A(\tau).$$

Proof. Given $x \in K(X)$, let $x = f^K(y)$ with $y \in K(Y)$. Then

$$\begin{aligned}
\rho f_K(x) &= \rho f_K f^K(y) \\
&= \rho(f_K(1)y) && \text{by projection formula} \\
&= \rho(f_K(1))\rho(y) \\
&= f_A(1)\tau\rho(y) && \text{by assumption} \\
&= f_A(f^A(\tau\rho(y))) && \text{by projection formula} \\
&= f_A(f^A(\tau)f^A\rho(y)) \\
&= f_A(f^A(\tau)\rho f^K(y)) \\
&= f_A(\tau_f\, \rho(x)),
\end{aligned}$$

as required.

The next notion is an abstract version of the main properties of deformations which will be constructed in Chapter IV, §5. Let

$$f: X \to Y$$

be a morphism. We shall say that f admits a **basic deformation** to a morphism $f': X \to Y'$ **with respect to** (K, ρ, A) if there exist morphisms as shown in the following diagram:

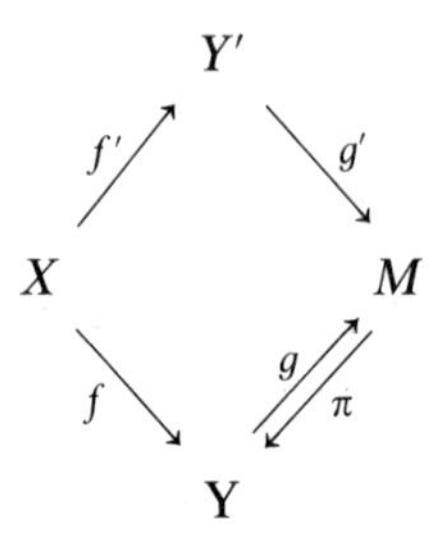

and a finite number of morphisms $h_\nu: C_\nu \to M$ with integers $m_\nu \in \mathbf{Z}$ satisfying the following conditions:

BD 1. *For each* $x \in K(X)$ *there exists some* $z \in K(M)$ *such that*

$$f_K(x) = g^K(z) \qquad \textit{and} \qquad f'_K(x) = g'^K(z).$$

BD 2. $g_A(1) = g'_A(1) + \sum m_\nu h_{\nu A}(1)$.

BD 3. *For each* $z \in K(M)$ *as in* **BD 1** *and all* ν, $h_\nu^K(z) = 0$.

BD 4. g *is a section of* π, *and* $\pi \circ g' \circ f' = f$.

Theorem 1.3. *Let* $f: X \to Y$ *be a morphism which admits a basic deformation to a morphism* f' *for which Riemann-Roch holds. Then Riemann-Roch holds for* f *with multiplier* $\tau_f = \tau_{f'}$.

Proof. Given $x \in K(X)$, choose z in $K(M)$ as in **BD 1**. Then

$$\begin{aligned}
g_A \rho f_K(x) &= g_A \rho g^K(z) && \text{by } \mathbf{BD\ 1}\\
&= g_A g^A \rho(z) && \text{by contravariance of } \rho\\
&= g_A(1)\rho(z) && \text{by projection formula}\\
&= g'_A(1)\rho(z) + \textstyle\sum m_\nu h_{\nu A}(1)\rho(z) && \text{by } \mathbf{BD\ 2}\\
&= g'_A g'^A \rho(z) + \textstyle\sum m_\nu h_{\nu A} h_\nu^A \rho(z) && \text{by projection formula}\\
&= g'_A \rho g'^K(z) + \textstyle\sum m_\nu h_{\nu A} \rho h_\nu^K(z) && \text{by contravariance of } \rho\\
&= g'_A \rho g'^K(z) && \text{by } \mathbf{BD\ 3}\\
&= g'_A \rho f'_K(x) && \text{by } \mathbf{BD\ 1}.
\end{aligned}$$

We now apply π_A. Since g is a section of π,

$$
\begin{aligned}
\rho f_K(x) &= \pi_A g_A \rho f_K(x) \\
&= \pi_A g'_A \rho f'_K(x) && \text{by the preceding steps} \\
&= \pi_A g'_A f'_A(\tau_{f'} \rho(x)) && \text{by R–R for } f' \\
&= f_A(\tau_{f'}(\rho(x))) && \text{by } \mathbf{BD\ 4}.
\end{aligned}
$$

This concludes the proof.

The reader may easily verify the following proposition.

Proposition 1.4. *If (K, ρ, L) and (L, σ, A) are Riemann–Roch functors, then $(K, \sigma\rho, A)$ is also a Riemann–Roch functor. If Riemann–Roch holds for f with respect to (K, ρ, L) (resp. (L, σ, A)) with multiplier τ_f (resp. ν_f), then Riemann–Roch holds for $(K, \sigma\rho, A)$ with multiplier $\nu_f\, \sigma(\tau_f)$.*

A Riemann–Roch functor can be obtained in the context of Chern classes as follows. A **Chern class functor** on $\mathfrak{C}$ is a triple (K, c, A), with K, A functors satisfying **F 1** to **F 3** and for each X in $\mathfrak{C}$ a Chern class homomorphism

$$c_X : K(X) \to 1 + A(X)^+$$

satisfying the following conditions:

CCF 1. *Each $K(X)$ is a λ-ring with involution, and f^K is a homomorphism of λ-rings with involution.*

CCF 2. *Each $A(X)$ is a graded ring, and f^A is a graded ring homomorphism of degree 0.*

CCF 3. *For $f : X \to Y$, $y \in K(Y)$, we have*

$$f^A c(y) = c(f^K(y)).$$

Since f^A and f^K are ring homomorphisms, when A is a $\mathbf{Q}$-algebra it *follows* that we also have the functorial rules

$$f^A \operatorname{ch}(y) = \operatorname{ch}(f^K(y)) \qquad \text{and} \qquad f^A \operatorname{td}(y) = \operatorname{td}(f^K(y)).$$

We conclude:

If $X \mapsto (K(X), c_X, A(X))$ is a Chern class functor, then

$$X \mapsto (K(X), \operatorname{ch}_X, \mathbf{Q}A(X))$$

is a Riemann–Roch functor.

On the other hand, we get a Riemann–Roch functor in a somewhat simpler situation as follows.

Let K be a functor from $\mathfrak{C}$ to λ-rings, satisfying **F 1**, **F 2**, **F 3** *and* **CCF 1**. *Then for each $j \geqq 0$ the Adams character*

$$\psi_X^j = \psi^j : K(X) \to K(X)$$

commutes with f^K, and therefore

$$(K, \psi^j, K) \text{ is a Riemann–Roch functor.}$$

Such functors K will be called λ-ring functors and will be studied in §3.

II §2. Grothendieck–Riemann–Roch for Elementary Imbeddings and Projections

We say that a morphism $f: X \to Y$ is an **elementary imbedding** with respect to the Chern class functor (K, c, A) if

$$f^K: K(Y) \to K(X)$$

is surjective, and there is a positive element q in $K(Y)$ such that

$$f_K(1) = \lambda_{-1}(q) \qquad \text{and} \qquad f_A(1) = c^{\mathrm{top}}(q^\vee).$$

Note that f^K is surjective whenever f is a section, i.e. there is a morphism π from Y to X with $\pi \circ f = \mathrm{id}_X$. The element q is called a **principal element** for the imbedding.

Consider the associated Riemann–Roch functor (K, ch, A), assuming A is a $\mathbf{Q}$-algebra.

Theorem 2.1. *Riemann–Roch holds for elementary imbeddings, with multiplier*

$$\tau_f = \mathrm{td}(f^K q^\vee)^{-1}.$$

Proof. This follows immediately from Theorem 1.2, and Proposition 5.3 of Chapter I.

Next we shall consider a "dual" situation. A morphism $f: X \to Y$ will be called an **elementary projection with respect to** (K, c, A) if the corresponding map

$$f_K: K(X) \to K(Y)$$

is isomorphic to the functional

$$f_e: K_e \to K$$

associated with some positive element e of $K = K(Y)$, and furthermore, letting $c = c(e)$, if

$$f_A: A(X) \to A(Y)$$

is isomorphic to the functional

$$g_c: A_c \to A.$$

These functionals were defined in §1 and §3 of Chapter I, respectively. By this we mean that there are two commutative diagrams

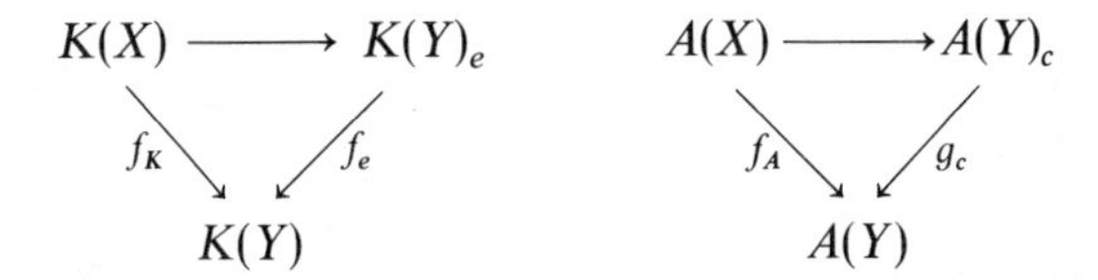

and that the top arrows are $K(Y)$ (resp. $A(Y)$-) isomorphisms, viewing $K(X)$ as a $K(Y)$-algebra via f^K, $A(X)$ as an $A(Y)$-algebra via f^A. Furthermore, under the identifications given by these isomorphisms, we require that

$$c^1(\ell) = w.$$

where ℓ and w are the canonical generators of K_e and A_c.

Theorem 2.2. *Riemann–Roch holds for elementary projections f, with multiplier*

$$\tau_f = \mathrm{td}(\ell e^\vee).$$

Proof. Let $\varepsilon(e) = r + 1$. Since K_e is generated as a K-algebra by the elements ℓ^k, $-r \leq k \leq 0$, and f_e and g_c are linear over K and A, it suffices to show that

$$\mathrm{ch}\, f_e(\ell^k) = g_c(\mathrm{td}(\ell e^\vee) \cdot \mathrm{ch}(\ell^k)).$$

On the left, we have

$$\mathrm{ch}\, f_e(\ell^k) = \begin{cases} 1 & \text{if } k = 0, \\ 0 & \text{if } -r \leq k < 0. \end{cases}$$

On the right, we have

$$g_c(\mathrm{td}(\ell e^\vee) \cdot \exp(kw)).$$

The formula to be proved is therefore a formal identity. Let $x_1, \ldots, x_{r+1}$ be Chern roots for e. Then $w - x_i$ are Chern roots for $\ell e^\vee$, so

$$\mathrm{td}(\ell e^\vee) = \prod_{i=1}^{r+1} \beta(w - x_i),$$

where β is the defining power series for the Todd class. The assertion therefore follows from the following purely algebraic lemma.

Lemma 2.3. *Let $x_1, \ldots, x_{r+1}$ be independent variables, and consider the ring*

$$\mathbf{Q}[[x_1, \ldots, x_{r+1}]][T] \Big/ \prod_{i=1}^{r+1} (T - x_i).$$

The class of the power series

$$e^{kT} \prod_{i=1}^{r+1} \beta(T - x_i)$$

in this ring is represented by a uniquely determined polynomial in T of degree $\leq r$ with coefficients in $\mathbf{Q}[[x_1, \ldots, x_{r+1}]]$. Then the coefficient of T^r in this polynomial is 0 for $-r \leqq k < 0$ and is 1 for $k = 0$.

Proof. The following proof is due to Roger Howe. We first begin by standard considerations concerning polynomials.

Let R be a ring without divisor of 0. Let

$$g(T) = T^d - a_{d-1}T^{d-1} - \cdots - a_0$$

by a polynomial in $R[T]$, and consider the equation

$$T^d = a_0 + a_1 T + \cdots + a_{d-1}T^{d-1}.$$

Let x be a root of $g(T)$. Then by using the Euclidean algorithm or induction, we see that for any integer $n \geqq d$ there is a relation

$$T^n = a_{0,n} + a_{1,n} + \cdots + a_{d-1,n}T^{d-1}$$

with coefficients $a_{ij} \in R$. If $F(T) \in R[T]$ is a polynomial, then we can write

$$F(x) = a_0(F) + a_1(F)x + \cdots + a_{d-1}(F)x^{d-1},$$

where the coefficients $a_i(F)$ lie in R and depend linearly on F. Suppose that the equation has d roots and that there is a factorization

$$g(T) = \prod_{i=1}^{d} (T - x_i)$$

Substituting x_i for x with $i = 1,\ldots,d$ and using Cramer's rule on the resulting system of linear equations yields

$$\Delta a_j(F) = W_j(F),$$

where Δ is the Vandermonde determinant, and $\Delta_j(F)$ is obtained by replacing the j-th column by ${}^t(F(x_1),\ldots,F(x_d))$, so

$$\Delta_j(F) = \begin{vmatrix} 1 & x_1 & \ldots & F(x_1) & \ldots & x_1^{d-1} \\ 1 & x_2 & \ldots & F(x_2) & \ldots & x_2^{d-1} \\ \vdots & \vdots & & \vdots & & \vdots \\ 1 & x_d & \ldots & F(x_d) & \ldots & x_d^{d-1} \end{vmatrix}.$$

If $\Delta \neq 0$ then

$$a_j(F) = \Delta_j(F)/\Delta.$$

If $F(T)$ is a power series in $R[[T]]$ and if R is a complete local ring, with $x_1,\ldots,x_d$ in the maximal ideal, and $x = x_i$ for some i, then we can evaluate $F(x)$ because the series converges. The above formula for the coefficients $a_j(F)$ remains valid.

Let $x_1,\ldots,x_d$ be independent variables, and let A be the ring

$$\mathbf{Q}[[x_1,\ldots,x_d]][T] \Big/ \prod_{i=1}^{d} (T - x_i).$$

Substituting some x_i for T induces a natural homomorphism φ_i of A onto

$$\mathbf{Q}[[x_1,\ldots,x_d]] = R.$$

and the map $\varphi \mapsto (\varphi_1(z),\ldots,\varphi_d(z))$ gives an imbedding of A into the product of R with itself d times.

Now we let $F(T)$ be the power series of Lemma 2.3, that is

$$F(T) = \mathbf{e}^{kT} \prod_{i=1}^{d} \frac{(T - x_i)\mathbf{e}^{T - x_i}}{\mathbf{e}^{T - x_i} - 1} = \mathbf{e}^{kT} \prod_{i=1}^{d} \beta(T - x_i).$$

Under the substitution of some x_j for T it becomes a power series in x_j and $x_j - x_i$, and thus converges in $\mathbf{Q}[[x_1,\ldots,x_d]]$.

Then we obtain

$$F(T) \equiv a_0(F) + \cdots + a_{d-1}(F)T^{d-1} \qquad \operatorname{mod} \prod_{i=1}^{d} (T - x_i),$$

with $a_0(F),\ldots,a_d(F) \in \mathbf{Q}[[x_1,\ldots,x_d]]$, and these coefficients are given by the formula with the determinants as above.

This follows for the power series simply by taking limits of polynomials formally converging to the power series.

We now come to the heart of the proof which computes the last coefficient using the expression in terms of the determinants. Let $\Delta = V(x_1,\ldots,x_d)$ where V denotes Vandermonde. We have

$$V(x_1,\ldots,x_d)a_{d-1}(F) = \begin{vmatrix} 1 & x_1 & \ldots & x_1^{d-2} & F(x_1) \\ 1 & x_2 & \ldots & x_2^{d-2} & F(x_2) \\ \vdots & \vdots & & \vdots & \vdots \\ 1 & x_d & \ldots & x_d^{d-2} & F(x_d) \end{vmatrix}.$$

Furthermore,

$$F(x_j) = \mathbf{e}^{kx_j} \prod_{n \neq j} \frac{(x_j - x_n)\mathbf{e}^{x_j - x_n}}{\mathbf{e}^{x_j - x_n} - 1}.$$

We use the inductive relation of the Vandermonde determinants

$$V(x_1,\ldots,x_d) = V(x_1,\ldots,\widehat{x}_j,\ldots,x_d)(-1)^{d-j} \prod_{n \neq j} (x_j - x_n).$$

We expand the determinant for $a_{d-1}(F)$ according to the last column, to get

$$a_{d-1}(F) = \sum_{j=1}^{d} \mathbf{e}^{(k+d-1)x_j} \prod_{n \neq j} \frac{1}{\mathbf{e}^{x_j} - \mathbf{e}^{x_n}}.$$

We use the inductive relation, and replace x_i by $\mathbf{e}^{x_i}$, which we denote by y_i for typographical reasons. We then get

$$V(y_1,\ldots,y_d)a_{d-1}(F) = \begin{vmatrix} 1 & y_1 & \ldots & y_1^{d-2} & y_1^{k+d-1} \\ 1 & y_2 & \ldots & y_2^{d-2} & y_2^{k+d-1} \\ \vdots & \vdots & & \vdots & \vdots \\ 1 & y_d & \ldots & y_d^{d-2} & y_d^{k+d-1} \end{vmatrix}.$$

If $k \neq 0$ then two columns on the right are the same, so the determinant is equal to 0. If $k = 0$ then we get the Vandermonde determinant on the right, so $a_{d-1}(F) = 1$. This concludes the proof of the lemma.

Summary. The results of this and the preceding section imply that, to prove Riemann–Roch for a morphism f with respect to (K, ch, A) it suffices to factor f into a composite $p \circ i$, where p is an elementary projection, and i admits a basic deformation to an elementary imbedding.

II §3. Adams Riemann–Roch for Elementary Imbeddings and Projections

In certain contexts of a Riemann–Roch theorem, the rings K and A are the same. Or we may start with a functor K and obtain from it in a natural way various Riemann–Roch functors as we shall see in Chapter III, §4 involving both cases when $K = A$ and $K \neq A$ affecting each other. Thus we have to make an appropriate definition. We shall say that a functor K from a category $\mathfrak{C}$ to λ-rings with involutions is a **λ-ring functor** if it satisfies axioms **F 1–F 3** and **CCF 1**, that is to each morphism $f: X \to Y$ in $\mathfrak{C}$ there are homomorphisms of abelian groups

$$f^K: K(Y) \to K(X) \qquad \text{and} \qquad f_K: K(X) \to K(Y)$$

such that:

$X \mapsto K(X)$ is a contravariant functor of λ-rings with involution via f^K;

$X \mapsto K(X)$ is a covariant functor of abelian groups via f_K;

the projection formula holds, that is for all morphisms $f: X \to Y$,

$$f_K(xf^K(y)) = f_K(x)y, \qquad \text{all } x \in K(X),\ y \in K(Y).$$

Except in Chapter VI, in this book our functors are both covariant and contravariant. In a context where singly variant functors occur as well, one might add the qualification that the above three properties define a **doubly variant** λ-ring functor.

Throughout this section we let K be a λ-ring functor as above, so that we have Riemann–Roch functors (K, ψ^j, K) with integers $j \geqq 0$ as mentioned at the end of §1. In all λ-rings arising in the sequel ($K(X)$, $K(X)_e$, etc.,) if u is a line element we assume that $1 - u$ is nilpotent.

A morphism $f: X \to Y$ is called an **elementary imbedding with respect to K** if

$$f^K: K(Y) \to K(X)$$

is surjective, and there is a positive element $e \in K(Y)$ such that

$$f_K(1) = \lambda_{-1}(e).$$

Remark. The surjectivity in practice comes from the fact that f is a section of a morphism $Y \to X$. The additional property of a section plays no role here, but will play a role in Chapter VI, §1 and §2.

Theorem 3.1. *Riemann-Roch holds for elementary imbeddings, with respect to* (K, ψ^j, K), *with multiplier*

$$\theta^j(f^K(e)).$$

Proof. This follows from Theorem 1.2, and Chapter I, Proposition 6.2.

A morphism $f: X \to Y$ is called an **elementary projection with respect to** K if the corresponding map

$$f_K: K(X) \to K(Y)$$

is isomorphic to the functional

$$f_e: K_e \to K$$

associated with some positive element e in $K = K(Y)$.

Theorem 3.2. *Let* f *be an elementary projection. If* j *is invertible in* $K = K(Y)$ *then* $\theta^j(e\ell^\vee)$ *is invertible in* $K_e = K(X)$, *and Riemann-Roch holds for* f *with respect to* (K, ψ^j, K), *with multiplier*

$$\tau_f = j\theta^j(e\ell^\vee)^{-1}.$$

Proof. We shall reduce the theorem to a formal identity of power series, similar to that of Theorem 2.2, over any ring where j is invertible. However, there is an alternative proof as follows. Since the identity is formal, one can verify it when K is replaced by $\mathbf{Q} \otimes K$. In this case, we shall prove in Theorem 4.3 of Chapter III (applied to the element $q = -e\ell^\vee + 1$) that Theorem 3.2 is actually a consequence of Theorem 2.2.

As to the power series proof, let us begin with the invertibility of θ^j. If u is a line element then

$$1 + u + \cdots + u^{j-1}$$

is invertible, because we can write $u = (u - 1) + 1$ and use the nilpotence of $u - 1$ and the geometric series to do the inversion. Let

$$e = \sum_{i=1}^{d} u_i$$

be a splitting of e. We have by definition

$$\theta^j(e) = \prod_{i=1}^{d} (1 + u_i + \cdots + u_i^{j-1}),$$

so $\theta^j(e)$ is invertible, and also $\theta^j(e\ell^\vee)$ is invertible. Recall

$$\psi^j(u) = u^j \qquad \text{for any line element } u.$$

We must show that the following diagram commutes:

$$\begin{array}{ccc} K_e & \xrightarrow{j\theta^j(e\ell^{-1})^{-1}\psi^j} & K_e \\ {\scriptstyle f_e}\downarrow & & \downarrow{\scriptstyle f_e} \\ K & \xrightarrow[\psi^j]{} & K \end{array}$$

i.e. show that

$$jf_e(\theta^j(e\ell^{-1})^{-1}\psi^j(x)) = \psi^j(f_e(x)) \qquad \text{for} \quad x \in K_e.$$

But f_e is K-linear, and ψ^j is a ring homomorphism, so it suffices to prove this commutativity relation for the elements $x = \ell^{-n}$, $0 \leqq n \leqq d - 1$, which form a basis of K_e over K. Recall that

$$f_e(\ell^{-n}) = \begin{cases} 1 & \text{if } n = 0, \\ 0 & \text{if } 1 \leqq n \leqq d - 1. \end{cases}$$

The desired commutativity amounts to proving

$$f_e(\theta^j(e\ell^{-1})^{-1}\ell^{-nj}) = \begin{cases} 1/j & \text{if } n = 0, \\ 0 & \text{if } n = 1, \ldots, d - 1. \end{cases}$$

At this point it is useful to adopt a notation which is used in the theory of formal groups, but the reader does not need to know any part of this theory. We begin with some general comments.

Let A be a commutative ring and I an ideal such that every element of I is nilpotent, or A is complete in the I-adic topology. We also assume that j is invertible in A. If $a \in I$ then $1 + a$ is invertible in A. For $a, b \in I$ we define

$$a[+]b = (1 + a)(1 + b) - 1 \quad \text{and} \quad a[-]b = (1 + a)(1 + b)^{-1} - 1.$$

The law of addition $[+]$ is associative and commutative, and makes I into a group, where 0 is the additive zero element. This group is called the **formal multiplicative group**. For our purposes, we don't need to know anything besides the definition of $[+]$ and $[-]$. We also use the notation of "variables", so if we denote by Z the "variable" of this group, then we can write

$$Z[+]Z' = (1 + Z)(1 + Z') - 1.$$

For any positive integer j we have the operator $[j]$ which is defined by iteration:

$$[j]Z = Z[+]Z[+]\cdots[+]Z \qquad \text{taken } j \text{ times.}$$

Then $[j]Z \equiv jZ \bmod Z^2$.

We let $a_i = u_i - 1$, and for $n = 0, \ldots, d - 1$ we let

$$F_n(Z) = (1 + Z)^{nj} \prod_{i=1}^{d} \frac{Z[+]a_i}{[j](Z[+]a_i)}.$$

Then $F_n(Z)$ is symmetric in the a_i, and is in fact a power series with coefficients in $K[[Z]]$ since we assumed j invertible in K. Furthermore

$$\theta^j(e\ell^{-1})^{-1}\ell^{-nj} = F_n(\ell^{-1} - 1).$$

The formula for F_n is of course motivated by the formal expression

$$\theta^j(e\ell^{-1})^{-1} = \prod_{i=1}^{d} \frac{u_i\ell^{-1} - 1}{(u_i\ell^{-1})^j - 1},$$

which does not make sense, but which is useful nevertheless. The Riemann-Roch formula is being reduced to the following result on formal power series.

Lemma 3.3. *Let R be a commutative ring in which j is invertible. Let $a_1, \ldots, a_d$, Z be independent variables. Let $F_n(Z)$ be defined by the above product, so that*

$$F_n(Z) \in R[[a_1, \ldots, a_d, Z]].$$

There exist unique elements $b_0^{(n)}, \ldots, b_d^{(n)} \in R[[s_s, \ldots, s_d]]$ *(where* $s_1, \ldots, s_d$ *are the elementary symmetric functions of* $a_1, \ldots, a_d$*) such that*

$$F_n(Z) \equiv b_0^{(n)} + \cdots + b_{d-1}^{(n)} Z^{d-1} \mod \prod_{i=1}^{d} Z\,[+]\,a_i.$$

and we have

$$\sum_{\nu=0}^{d-1} (-1)^\nu b_\nu^{(n)} = \begin{cases} 1/j & \text{if } n = 0, \\ 0 & \text{if } n = 1, \ldots, d-1. \end{cases}$$

Indeed, the leading coefficient of $\prod Z\,[+]\,a_i$ is $\prod (1 + a_i)$ and is therefore invertible in $R[[a_1, \ldots, a_d]]$. The division algorithm applies to give the desired congruence and the uniqueness of the coefficients $b_0^{(n)}, \ldots, b_{d-1}^{(n)}$. We specialize to the case when $a_i = u_i - 1$ in K. We view $\ell^{-1} - 1$ as the generic root of $\prod Z\,[+]\,a_i$ in

$$K[Z]/\prod Z\,[+]\,a_i.$$

Since $f_e(\ell^{-1}) = 0$, substituting $\ell^{-1} - 1$ for Z, and then -1 for $\ell^{-1} - 1$ amounts to substituting -1 for Z in the polynomial

$$b_0^{(n)} + \cdots + b_{d-1}^{(n)} Z^{d-1}.$$

This accomplishes the desired reduction to the formal power series relation of Lemma 3.3.

Furthermore, it suffices to prove the relation of Lemma 3.3 when $R = \mathbf{Z}[1/j]$ because in fact the coefficients $b_0^{(n)}, \ldots, b_{d-1}^{(n)}$ lie in the image of $\mathbf{Z}[1/j][[a_1, \ldots, a_d]]$ and are determined by universal formulas, which we shall make explicit with the Vandermonde determinants, as in Lemma 2.3. So we let $R = \mathbf{Z}[1/j]$.

The polynomial $\prod Z\,[+]\,a_i$ has d roots

$$z_i = [-1]a_i \quad \text{in} \quad R[[a_1, \ldots, a_d]].$$

Let

$\Delta = V(z_1, \ldots, z_d) =$ Vandermonde determinant;

$\Delta_\nu(F_n) =$ determinant obtained by replacing the ν-th column in the Vandermonde determinant by ${}^t(F_n(z_1), \ldots, F_n(z_d))$, so

$$\Delta_\nu(F_n) = \begin{vmatrix} 1 & z_1 & \cdots & F_n(z_1) & \cdots & z_1^{d-1} \\ 1 & z_2 & \cdots & F_n(z_2) & \cdots & z_2^{d-1} \\ \vdots & \vdots & & \vdots & & \vdots \\ 1 & z_d & \cdots & F_n(z_d) & \cdots & z_d^{d-1} \end{vmatrix}.$$

Then

$$\Delta b_\nu^{(n)} = \Delta_\nu(F_n)$$

and

$$\Delta \sum_{\nu=0}^{d-1} (-1)^\nu b_\nu^{(n)} = \sum_{\nu=0}^{d-1} (-1)^\nu \Delta_\nu(F_n).$$

$$= \sum_{\nu=0}^{d-1} \sum_{k=1}^{d} (-1)^{k+1} F_n(z_k) M_{k\nu}(z_1,\ldots,z_d),$$

where $M_{k\nu}(z_1,\ldots,z_d)$ is obtained by deleting the k-th row and ν-th column from Vandermonde, that is the $k\nu$-minor of the Vandermonde determinant. We invert the order of summation and use the next lemma.

Lemma 3.4.

$$\sum_{\nu=0}^{d-1} M_{k\nu}(z_1,\ldots,z_d) = V(z_1,\ldots,\hat{z}_k,\ldots,z_d) \prod_{i\neq k} (1 + z_i).$$

Proof. This is a special case of the Jacobi–Trudy identities, cf. [F 2], Lemma A.9.3, p. 422 which contains a short proof of the general identities. One can evaluate each term in the sum, namely

$$M_{k\nu}(z_1,\ldots,z_d) = V(z_1,\ldots,\hat{z}_k,\ldots,z_d) s_{d-1-\mu}(z_1,\ldots,\hat{z}_k,\ldots,z_d),$$

where $s_{d-1-\mu}$ is the elementary symmetric function. The index μ is used to index consecutively the variables with z_k omitted, so $\mu = 1,\ldots,d-1$. The general identities give the value of a perturbation of the Vandermonde determinant, whereby the powers of the variables are increased by certain amounts, denoted by λ_i in the above reference. Here the amount is 1 for all the powers from the ν-th column onward. This shows how the lemma is a special case of the Jacobi–Trudy identities.

Going on with the proof, since $[j]Z \equiv jZ \bmod Z^2$ we find

$$F_n(z_k) = F_n([-1]a_k) = (1 + z_k)^{nj} \frac{1}{j} \prod_{i\neq k} \frac{a_i[-]a_k}{[j](a_i[-]a_k)}$$

$$= (1 + z_k)^{nj} \frac{1}{j} \prod_{i\neq k} \frac{u_i - u_k}{u_k} \frac{u_k^j}{u_i^j - u_k^j},$$

where $u_i = 1 + a_i$. But $1 + z_i = u_i^{-1}$, so by Lemma 3.4:

$$\Delta \sum_{\nu=0}^{d-1} (-1)^\nu b_\nu^{(n)}$$

$$= \sum_{k=1}^{d} (-1)^{k+1} u_k^{-nj} V(z_1,\ldots,\hat{z}_k,\ldots,z_d) \frac{1}{j} \prod_{i\neq k} \frac{u_i - u_k}{u_k} \frac{u_k^j}{u_i^j - u_k^j} \frac{1}{u_i}.$$

We now use the recursive product for Vandermonde:

$$V(T_1,\ldots,T_d) = V(T_1,\ldots,\hat{T}_k,\ldots,T_d)(-1)^{d-k}\prod_{i\neq k}(T_k - T_i).$$

On the one hand, we have $z_k - z_i = (u_i - u_k)/u_k u_i$, so that

$$V(z_1,\ldots,\hat{z}_k,\ldots,z_d) = (-1)^{d-k}V(z_1,\ldots,z_d)\prod_{i\neq k}\frac{u_i u_k}{u_i - u_k}.$$

On the other hand,

$$\prod_{i\neq k}\frac{1}{u_i^j - u_k^j} = (-1)^{d-1+d-k}\frac{V(u_1^j,\ldots,\widehat{u_k^j},\ldots,u_d^j)}{V(u_1^j,\ldots,u_d^j)}.$$

Therefore

$$\Delta\sum(-1)^\nu b_\nu^{(n)}$$

$$= \frac{1}{j}\frac{V(z_1,\ldots,z_d)}{V(u_1^j,\ldots,u_d^j)}\sum_{k=1}^{d}(-1)^{d-k}u_k^{-nj}u_k^{j(d-1)}V(u_1^j,\ldots,\widehat{u_k^j},\ldots,u_d^j)$$

If $n = 1,\ldots,d-1$ then this last expression is 0 because it is the expansion of a determinant with two equal columns. If $n = 0$, then the sum on the right is the expansion of $V(u_1^j,\ldots,u_d^j)$ according to the last column, so we find the value

$$\frac{1}{j}\Delta$$

thereby proving Lemma 3.3 and also Theorem 3.2.

II §4. An Integral Riemann-Roch Formula

Although the formalism developed in §1 and §2 was based on having a *ring* homomorphism ρ from K to A, some of the same ideas can be used in other contexts. We illustrate this by a "Grothendieck-Riemann-Roch theorem without denominators", which can be used to compute Chern classes, and not just the Chern character (which requires denominators). Such a formula was first given by Grothendieck, and proved more generally by Jouanolou [J], cf. [BFM 1], [F 2].

First we establish systematically another general formula for the Chern classes. If

$$e = \sum_{i=1}^{r} u_i$$

is a splitting of a positive element e, and $a_i = c^1(u_i)$, then

$$c_t(e) = \prod_{i=1}^{r} (1 + a_i t)$$

is a splitting of the Chern polynomial c_t. We shall be concerned with the Chern polynomials of various combinations of positive elements. As we saw in §1,

$$c_t(\lambda^j(e)) = \prod_{k_1 < \cdots < k_j} (1 + (a_{k_1} + \cdots + a_{k_j})t).$$

We wish to compute the Chern class of a combination

$$c(\lambda_{-1}(q^\vee)e)$$

instead of $\mathrm{ch}(\lambda_{-1}(q^\vee)e)$ which would introduce denominators.

For the moment, let $a_1,\ldots,a_r$ and $b_1,\ldots,b_d$ be independent variables. We define the power series

$$Q_{r,d}(a_1,\ldots,a_r; b_1,\ldots,b_d)$$

to be the power series with integer coefficients given by the formula

$$Q_{r,d}(a, b) = \prod_{i=1}^{r} \prod_{j=0}^{d} \prod_{k_1 < \cdots < k_j} (1 + a_i - b_{k_1} - \cdots - b_{k_j})^{(-1)^j}.$$

This power series depends only on the integers d and r and has constant term equal to 1. It is symmetric in $a_1,\ldots,a_r$ and also in $b_1,\ldots,b_d$. Therefore it can be expressed uniquely as a power series with integral coefficients in the symmetric functions

$$s_1(a),\ldots,s_r(a) \qquad \text{and} \qquad s_1(b),\ldots,s_d(b).$$

In addition

$$Q(a, b) - 1 \quad \textit{is divisible by the product} \quad b_1 \cdots b_d.$$

Proof. It suffices to show that $Q - 1$ is divisible by b_1, since $Q - 1$ is symmetric in $b_1,\ldots,b_d$. If we set $b_1 = 0$ then each term in the product with a given value of j and $k_1 > 1$ cancels a term with $k_1 = 1$ and j replaced by $j + 1$. This proves our assertion.

In light of the divisibility, there exists a unique power series $P^{\mathrm{sp}}_{r,d}(a, b)$ ("sp" for "split") with integral coefficients, such that

$$Q_{r,d}(a, b) = 1 + b_1 \cdots b_d P^{\mathrm{sp}}_{r,d}(a, b).$$

Again, $P^{\mathrm{sp}}_{r,d}(a, b)$ is symmetric in $a_1, \ldots, a_r$ and also in $b_1, \ldots, b_d$. Therefore there exists a unique power series $P_{r,d}$ with integer coefficients such that

$$P^{\mathrm{sp}}_{r,d}(a, b) = P_{r,d}(s_1(a), \ldots, s_r(a); s_1(b), \ldots, s_d(b)).$$

In the context of λ-rings and a Chern class homomorphism c, we can now substitute first Chern classes for the variables a, b. If e, q are positive elements of augmentation r, d, respectively, we shall use the notation

$$P_{r,d}(e, q) = P_{r,d}(c^1(e), \ldots, c^r(e); c^1(q), \ldots, c^d(q)),$$

where c^i are the Chern classes.

We recall that in the context of Chern classes, if $b_j = c^1(v_j)$, then

$$c^{\mathrm{top}}(q) = b_1 \cdots b_d.$$

The definition of $P_{r,d}$ has been made in such a way that by applying the formula for $c_t(\lambda^j e)$ we immediately find:

Proposition 4.1. *Let c be a Chern class homomorphism. For positive elements e, q of augmentations r, d respectively, we have*

$$c(\lambda_{-1}(q^\vee)e) = 1 + c^{\mathrm{top}}(q)P_{r,d}(e, q).$$

This formula now looks formally similar to the formula in Proposition 5.3, which led to a Riemann-Roch theorem for elementary imbeddings. Hence we are led to make the appropriate definitions in the present context. Let (K, c, A) be a Chern class functor. Let $f: X \to Y$ be a morphism. We shall say that **Integral Riemann-Roch** holds for f if there exists a positive element $q \in K(X)$ such that for any positive element e in $K(X)$ we have

$$c(f_K(e)) = 1 + f_A(P_{r,d}(e, q)),$$

where $d = \varepsilon(q)$ and $r = \varepsilon(e)$. We call q a **Riemann-Roch element** for f.

Theorem 4.2. *Integral Riemann-Roch holds if f is an elementary imbedding.*

Proof. This follows from Proposition 4.1, exactly as Theorem 2.1 followed from Proposition 5.2 of Chapter I.

Theorem 4.3. *Let f be a morphism which admits a basic deformation to a morphism f' for which Integral Riemann–Roch holds. Then Integral Riemann–Roch holds for f with the same Riemann–Roch element q.*

Proof. Identical with the proof of Theorem 1.3, replacing ρ by c; that proof did not require ρ to be a ring homomorphism.

CHAPTER III

Grothendieck Filtration and Graded K

The object of the first two sections is to construct from a λ-ring K a graded ring $\operatorname{Gr} K$, with a Chern class homomorphism satisfying the properties of Chapter I, §3.

III §1. The γ-Filtration

We let K be a λ-ring as in Chapter I, §1. Define the operations

$$\gamma^i : K \to K$$

by the series

$$\gamma_t(x) = \lambda_{t/(1-t)}(x) = \sum \lambda^i(x)\left(\frac{t}{1-t}\right)^i = \sum \gamma^i(x)t^i.$$

Since $t/(1-t) = s$ is another parameter generating the power series ring

$$K[[t]] = K[[s]],$$

we see that the γ^i also define a λ-ring structure on K: that is, for all positive integers k we have $\gamma^0(x) = 1$, $\gamma^1(x) = x$ and

$$\gamma^k(x + y) = \sum_{i=0}^{k} \gamma^i(x)\gamma^{k-i}(y).$$

In addition, it follows immediately from the definition that if $u \in \mathbf{L}$, then

$$\gamma_t(u - 1) = 1 + (u - 1)t \quad \text{and so} \quad \gamma^i(u - 1) = 0 \quad \text{for} \quad i > 1;$$

$$\gamma_t(1 - u) = \sum (1 - u)^i t^i \quad \text{and so} \quad \gamma^i(1 - u) = (1 - u)^i \quad \text{for} \quad i \geqq 0.$$

Proposition 1.1. *Let e be a positive element of K with $\varepsilon(e) = m$. Then*

(a)
$$\sum_{i=0}^{m} \gamma^i(e-m)t^{m-i} = \sum_{i=0}^{m} \lambda^i(e)(t-1)^{m-i}.$$

(b)
$$\gamma^m(e-m) = \lambda_{-1}(e^\vee)\cdot\lambda^m(e).$$

Proof. From the definition of γ, using a new variable v, we get

$$\begin{aligned}\gamma_v(e-m) &= \frac{\gamma_v(e)}{\gamma_v(m)} = \lambda_{v/(1-v)}(e)(1-v)^m \\ &= \sum_{i=0}^{m} \lambda^i(e)v^i(1-v)^{m-i}.\end{aligned}$$

Setting $v = t^{-1}$ and multiplying by t^m yields (a).

For (b), set $t = 0$ in (a) and use Lemma 5.1 of Chapter I:

$$\begin{aligned}\gamma^m(e-m) &= \sum_{i=0}^{m} \lambda^i(e)(-1)^{m-i} \\ &= \sum (-1)^{m-i}\lambda^{m-i}(e^\vee)\lambda^m(e) \\ &= \lambda_{-1}(e^\vee)\lambda^m(e).\end{aligned}$$

This proves the proposition.

Next we introduce the **Grothendieck γ-filtration**. We let

$$F^1 = F^1K = \operatorname{Ker}\varepsilon$$

Then for $n \geqq 1$ we let

$F^n = F^nK = \mathbf{Z}$-module generated by the elements $\gamma^{r_1}(x_1)\cdots\gamma^{r_k}(x_k)$ with

$$x_1,\ldots,x_k \in F^1 \qquad \text{and} \qquad \sum r_i \geqq n.$$

It is immediately verified that this defines a filtration, and F^n is an ideal for each n, because

$$x\gamma^{r_1}(x_1)\cdots\gamma^{r_k}(x_k) = (x-\varepsilon(x))\gamma^{r_1}(x_1)\cdots\gamma^{r_k}(x_k) + \varepsilon(x)(\cdots),$$

and the first term on the right-hand side belongs to F^{n+1}.

It is convenient to have the filtration defined for all integers, so we let

$$F^n = K \qquad \text{for} \quad n \leqq 0.$$

Since the augmentation gives a homomorphism of K onto $\mathbf{Z}$, we have a natural isomorphism

$$F^0/F^1 = K/\operatorname{Ker} \varepsilon \approx \mathbf{Z}.$$

We note that in case K is generated by line elements, then F^1 is generated over $\mathbf{Z}$ by elements $u - 1$ with $u \in \mathbf{L}$, and

$$F^i = (F^1)^i$$

for all $i \geqq 1$. Indeed

$$(u_1 - 1)\cdots(u_n - 1) = \gamma^1(u_1 - 1)\cdots\gamma^1(u_n - 1)$$

so $(F^1)^n \subset F^n$. Conversely, it suffices to prove that $\gamma^i(x) \in (F^1)^i$ for all $i \geqq 1$, and $x \in K$. From the values $\gamma^j(u - 1)$ and $\gamma^j(1 - u)$ which we derived previously, the desired inclusion follows at once.

We turn next to proving a

Graded Splitting Property. *Given a positive element $e \in K$, there exists a λ-ring extension K' (with involution if relevant) such that*

$$F^n K' \cap K = F^n K$$

for all integers $n \geqq 0$.

As in Chapter I, §2, we consider the extension

$$K_e = K[\ell],$$

where ℓ is the generic root of the equation

$$\sum_{i=0}^{\varepsilon(e)} (-1)^i \lambda^i(e) \ell^{\varepsilon(e)-i} = 0.$$

From Proposition 1.1(a), setting $t = 1 - \ell$, we see that $\ell - 1$ is the generic root of the equation

$$\boxed{\sum_{i=0}^{\varepsilon(e)} (-1)^i \gamma^i(e - \varepsilon(e))(\ell - 1)^{\varepsilon(e)-i} = 0.}$$

For present purposes it is convenient to let

$$\varepsilon(e) = r + 1.$$

By Chapter I, Theorem 2.1, K_e is a λ-ring extension of K, so K_e has a γ-filtration $F^n K_e$. We write $F^n = F^n K$. Recall that $F^n = K$, if $n \leqq 0$.

Theorem 1.2. *For all integers $k \geqq 0$ we have*

$$F^k K_e = \sum_{i=0}^{r} F^{k-i}(\ell - 1)^i.$$

Proof. Let $x = \ell - 1$. Define

$$R_k = \sum_{i=0}^{\infty} F^{k-i} \cdot x^i.$$

Note first that the R_k form a ring filtration of K_e, i.e.,

$$R_j \cdot R_k \subset R_{j+k}.$$

We need the following

(*) *If y, $z \in K_e$, and k is a positive integer such that $\gamma^i(y) \in R_i$ and $\gamma^i(z) \in R_i$ for all $1 \leqq i \leqq k$, then $\gamma^k(y \cdot z) \in R_k$.*

The statement follows from the existence of universal polynomials P_k of weight k such that

$$\gamma^k(y \cdot z) = P_k(\gamma^1(y), \ldots, \gamma^k(y), \gamma^1(z), \ldots, \gamma^k(z)).$$

Next we claim that

$$R_k = F^k K_e. \tag{1.1}$$

That R_k is contained in $F^k K_e$ follows from the fact that $\gamma^1 x = x$ and $\gamma^i x = 0$ for $i > 1$. For the reverse inclusion it suffices to show that if $y \in F^1 K_e$, then $\gamma^k y \in R_k$. Writing $y = \sum a_i x^i$, $a_i \in K$, then $\varepsilon(a_0) = \varepsilon(y) = 0$, so it suffices to show that for $a \in K$, $i > 0$, we have

$$\gamma^k(a x^i) \in R_k.$$

This follows, by induction on i, from (*).

From the equation for x we have

$$x^{r+1} \in F^{r+1} + F^r \cdot x + \cdots + F^1 \cdot x^r.$$

It follows by induction on j that

$$(**) \qquad x^j \in F^j + F^{j-1}\cdot x + \cdots + F^{j-r}\cdot x^r$$

for all $j > r$. For if this holds for j, then

$$x^{j+1} \in F^j\cdot x + F^{j-1}\cdot x^2 + \cdots + F^{j-r}\cdot x^{r+1}$$

and $F^{j-r}x^{r+1} \subset F^{j-r}\cdot(F^{r+1} + F^r\cdot x + \cdots + F^1\cdot x^r)$, so

$$x^{j+1} \in F^{j+1} + F^j\cdot x + \cdots + F^{j-r+1}\cdot x^r.$$

Finally we have the equalities

$$(1.2) \qquad R_k = \sum_{i=0}^{k+r+1} F^{k-i}\cdot x^i = \sum_{i=0}^{r} F^{k-i}\cdot x^i$$

The first equality follows from the equation $K_e = \sum_{j=0}^{r} F^0\cdot x^j$. The second follows from $(**)$, since for $i > r$,

$$\begin{aligned} F^{k-i}\cdot x^i &\subset F^{k-i}(F^i + F^{i-1}\cdot x + \cdots + F^{i-r}\cdot x^r) \\ &\subset F^k + F^{k-1}\cdot x + \cdots + F^{k-r}\cdot x^r. \end{aligned}$$

The theorem follows from (1.1) and (1.2).

Corollary 1.3. *Let $f_e\colon K_e \to K$ be the functional such that $f_e(\ell^i) = \sigma^i(e)$ for all $i \geqq 0$. Then for all k,*

$$f_e(F^k K_e) \subset F^{k-r}.$$

Proof. Immediate from Theorem 1.2 and the K-linearity of f.

It follows from Theorem 1.2 that

$$F^k = F^k(K_e) \cap K.$$

The graded splitting principle then follows as in the argument in Chapter I, §2 by constructing a chain of elementary extensions

$$K \subset K^{(1)} = K_e \subset \cdots \subset K^{(r)} = K'$$

so that e splits in K', and $F^k K' \cap K = K^k$.

In the applications, the elements of F^1 will be nilpotent. In fact, something much stronger will be proved in Chapter V, Corollary 3.10, namely that $F^i = 0$ for i sufficiently large. Here we give another proof of nilpotency, but *the rest of this section will not be used any further in the book.*

A line element $u \in \mathbf{L}$ will be called **ample** for K if, given $x \in K$ there is an integer $n(x)$ such that for all $n \geqq n(x)$,

$$u^n x = e - m$$

for some positive e and some integer m. (To see where this terminology comes from, see Chapter V, Lemma 3.1.)

Lemma 1.4. *If u is ample for K, then for any $v \in \mathbf{L}$, $v - 1$ is nilpotent.*

Proof. For $n \geqq n(-v^{-1})$ we may write

$$-v^{-1}u^n = e - m$$

for $e \in \mathbf{E}$, $m > 0$. Let $w = vu^{-n}$. Then

$$mw - 1 = (e + v^{-1}u^n)w - 1 = ew$$

lies in $\mathbf{E}$, so for a suitable positive integer k we have

$$\begin{aligned} 0 = \lambda^k(mw - 1) &= \sum_{i=0}^{k} (-1)^{k-i}\lambda^i(mw) \\ &= (-1)^k \lambda_{-1}(mw) \\ &= (-1)^k \lambda_{-1}(w)^m = (-1)^k(1 - w)^m. \end{aligned}$$

Thus $1 - w$ is nilpotent. The same argument, with $v = 1$, $w = u^{-n}$ shows that $1 - u^{-n}$ is nilpotent for sufficiently large n. Therefore

$$1 - v = (1 - w) - v(1 - u^{-n})$$

is also nilpotent.

Proposition 1.5. *Assume that for each positive e in K there is an extension K' satisfying the graded splitting property for e, and having an ample line element. Then every element of F^1K is nilpotent.*

Proof. Immediate from Lemma 1.4 and the splitting property.

We give an application of the splitting property.

Lemma 1.6. *Given an element $x \in F^n K$, there exists an extension K' of K such that x can be written as a linear combination with integer coefficients*

$$(u_1 - 1)^{m_1} \cdots (u_k - 1)^{m_k}$$

with line elements u_i and positive integers m_i such that

$$\sum m_i \geqq n.$$

Proof. This is a version for one element of a fact we have already noticed that $F^i = (F^1)^i$ if K is generated by line elements. One could also apply Zorn's lemma to splitting extensions to get a huge extension K' which satisfies this property.

Theorem 1.7. *Let $\mathbf{L}$ be the multiplicative group of line elements. Then the map $u \mapsto u - 1$ induces an isomorphism*

$$\mathbf{L} \xrightarrow{\approx} \mathrm{Gr}^1(K) = F^1K/F^2K.$$

Proof. The map is obviously a homomorphism into $\mathrm{Gr}^1(K)$. We shall construct an inverse. As usual, let $\mathbf{E}$ be the set of positive elements. Let

$$\det : \mathbf{E} \to \mathbf{L}$$

be the map such that

$$\det(e) = \lambda^r(e) \qquad \text{if} \quad \varepsilon(e) = r.$$

If $e = e' + e''$ then $\det(e' + e'') = \det(e')\det(e'')$ from the addition formula for $\lambda^r(e' + e'')$, combined with the fact that $\lambda^i(e') = 0$ if $i > \varepsilon(e')$ and similarly for e''. Hence det is a homomorphism of $\mathbf{E}$ into $\mathbf{L}$ which extends to a homomorphism of K into $\mathbf{L}$.

This map det is trivial on F^2K. To see this, let $x \in F^2K$. By the splitting principle, in some extension of K we can write x as a linear combination with integer coefficients of elements

$$(u_1 - 1)^{m_1} \cdots (u_k - 1)^{m_k}$$

with line elements u_i and positive integers m_i such that $\sum m_i \geqq 2$. Such an element contains some factor

$$(u - 1)(v - 1) = uv - v - u + 1,$$

and for any line element w, it is immediate that

$$w(u - 1)(v - 1) = wuv - wv - wu + w$$

lies in the kernel of det, so $F^2K \subset \operatorname{Ker} \det$ as asserted. Thus det induces a homomorphism

$$\det : K/F^2K \to \mathbf{L}.$$

Let $g: \mathbf{L} \to \operatorname{Gr}^1(K)$ be the homomorphism $u \mapsto u - 1 \bmod F^2K$. Since $\det(u - 1) = \det u = u$, it follows that $\det \circ g = \mathrm{id}$.

Conversely, to show that $g \circ \det = \mathrm{id}$ on $\operatorname{Gr}^1(K)$, we use the splitting principle. Let $x \in F^1K$ so $\varepsilon(x) = 0$. We can write

$$x = \sum n_i u_i = \sum n_i(u_i - 1)$$

with $n_i \in \mathbf{Z}$ and line elements u_i in an extension of K. Then

$$\det(x) = \prod u_i^{n_i}$$

and $g \circ \det(x) = x \bmod F^2K$. This concludes the proof.

For the interpretation in the geometric context, see the end of Chapter V, §3.

III §2. Graded K and Chern Classes

Associated with the filtration F^n on K, we have the **associated graded ring**

$$\operatorname{Gr}(K) = \bigoplus_{k=0}^{\infty} F^k/F^{k+1}.$$

When no confusion is likely we write G for $\operatorname{Gr}(K)$, and $G^k = \operatorname{Gr}^k K$ for the k-th graded piece F^k/F^{k+1}. For a positive element e in K define the i-th **Chern Class** to be

$$c^i(e) = \gamma^i(e - \varepsilon(e)) \quad \bmod F^{i+1},$$

so $c^i(e) \in G^i$. If $\varepsilon(e) = m$, and

$$e = \sum_{i=1}^{m} u_i$$

with $u_i \in \mathbf{L}$, then we put $a_i = c^1(u_i) = u_i - 1 \bmod F^2$. Therefore

$$c_t(e) = \sum c^i(e)t^i = \prod_{i=0}^{m} (1 + a_i t).$$

With the present definition of Chern classes, we see that the isomorphism $\mathbf{L} \to \mathrm{Gr}^1(K)$ of Theorem 1.7 is given by the first Chern class

$$u \mapsto c^1(u) = u - 1 \mod F^2.$$

The fact that γ_t defines a λ-ring structure on K implies that the map defined on $\mathbf{E}$ by

$$c_t : e \mapsto \sum c^i(e)t^i$$

and extended by additivity to K is a Chern class homomorphism in the sense of Chapter I, §3, provided $c^i(e)$ is nilpotent for $i > 0$. The splitting principle follows from the graded splitting principle of §1. Proposition 1.5 (using an ample element) or better Corollary 3.10 of Chapter V can be used to verify the axiom that the $c^i(e)$ are nilpotent. If we had not assumed **CC 3** and had taken values of Chern classes in $\hat{A}$ in Chapter I, §3 we would not need nilpotency. As it is:

For the rest of this section, we assume that all elements of Gr^k *are nilpotent for* $k \geqq 1$. *The same assumption is also made for* $\mathrm{Gr}^k K(X)$ *when* K *is a* λ*-ring functor below.*

A homomorphism $f^K : K \to K'$ of λ-rings maps $F^k K$ to $F^k K'$, so induces a homomorphism

$$f^G = \mathrm{Gr}(f^K) : \mathrm{Gr}\, K \to \mathrm{Gr}\, K'$$

of graded rings. This satisfies

$$f^G(c^i(x)) = c^i(f^K(x)) \qquad \text{for} \quad x \in K.$$

Suppose $f_K : K' \to K$ is a K-linear homomorphism via f^K. We say that f_K has **graded degree** d for some integer d if

$$f_K(F^k K') \subset F^{k+d} K$$

for all integers k. Then f_K induces a graded homomorphism

$$f_G = \mathrm{Gr}(f_K) : \mathrm{Gr}\, K' \to \mathrm{Gr}\, K$$

of degree d. Note that d may be negative. This map is $\mathrm{Gr}(K)$-linear, i.e. we have the projection formula

$$f_G(f^G(y)x) = y f_G(x).$$

The above discussion tells us how we shall obtain Riemann–Roch functors in practice:

Let K be a λ-ring functor satisfying the above nilpotency condition. Then with respect to all morphisms which have a graded degree, $(K, c, \mathrm{Gr}\, K)$ is a Chern class functor and $(K, \mathrm{ch}, \mathbf{Q}\mathrm{Gr}\, K)$ is a Riemann–Roch functor as in Chapter II, §1.

Here $\mathbf{Q}\mathrm{Gr}\, K$ denotes $\mathbf{Q} \otimes_{\mathbf{Z}} \mathrm{Gr}\, K$.

With these considerations, we may apply the Riemann–Roch formalism of Chapter II, taking the graded ring A to be $\mathrm{Gr}\, K$. We consider first elementary imbeddings, then elementary projections.

Proposition 2.1.

(i) *Let K be a λ-ring and $q \in K$ a positive element. Let $d = \varepsilon(q)$. Then*

$$c^{\mathrm{top}}(q^{\vee}) = \lambda_{-1}(q) \mod F^{d+1}.$$

(ii) *Let K be a λ-ring functor and let $f: X \to Y$ be a morphism such that $f^K: K(Y) \to K(X)$ is surjective and*

$$f_K(1) = \lambda_{-1}(q)$$

for some positive element $q \in K(Y)$. Let $d = \varepsilon(q)$. Then f has graded degree d, and

$$f_G(1) = c^{\mathrm{top}}(q^{\vee}).$$

Proof. By Proposition 1.1(b) we have

$$\gamma^d(q^{\vee} - d) = \lambda_{-1}(q)\lambda^d(q^{\vee}).$$

Since $\lambda^d(q^{\vee})$ is a line element, it follows that $\lambda^d(q^{\vee}) \equiv 1 \mod F^1$, so

$$c^{\mathrm{top}}(q^{\vee}) = \gamma^d(q^{\vee} - d) \equiv \lambda_{-1}(q) \mod F^{d+1}.$$

This proves the first assertion. Since $f^K: K(Y) \to K(X)$ is surjective and commutes with ε and γ^i, it follows from the definition of γ-filtration that f^K maps $F^m K(Y)$ onto $F^m K(X)$ for all m. Given $x \in F^m K(X)$, write

$$x = f^K(y), \qquad y \in F^m K(Y).$$

Then

$$f_K(x) = f_K(f^K y) = y \cdot f_K(1) = y \cdot \lambda_{-1}(q)$$

is in $F^{m+d}K(Y)$, so f has graded degree d. The value $f_G(1)$ comes from the first part of the proposition.

The conditions of Chapter II, §2 are therefore satisfied, and we have the

Corollary 2.2. *With the assumptions of the proposition, f is an elementary imbedding with respect to the Chern class functor $(K, c, \mathrm{Gr}\, K)$.*

In particular, Riemann–Roch holds for f with respect to $(K, \mathrm{ch}, \mathbf{Q}\mathrm{Gr}\, K)$, with multiplier

$$\tau_f = \mathrm{td}(f^K q^\vee)^{-1}.$$

For a positive element e in an arbitrary λ-ring K, consider the extension K_e of K. In Theorem 1.2, let $f^K: K \to K_e$ be the natural inclusion. (In the applications, f^K will arise from a λ-ring functor.) In light of Theorem 1.2, f^K induces an injective homomorphism on the graded rings, which we denote

$$f^G: \mathrm{Gr}(K) \to \mathrm{Gr}(K_e).$$

On the other hand, by Corollary 1.3, the functional $f_e: K_e \to K$ maps $F^k K_e$ to $F^{k-r}K$, where $\varepsilon(e) = r + 1$. Therefore f_e induces a homomorphism on the graded rings, which we denote

$$f_G: \mathrm{Gr}\, K_e \to \mathrm{Gr}\, K,$$

lowering degrees by r. Let:

$w \in \mathrm{Gr}^1 K_e$ denote the class of $\ell - 1 \bmod F^2 K_e$;

$$p_c(W) = \sum_{i=0}^{r+1} (-1)^i c^i(e) W^{r+1-i}.$$

If e splits into $\sum u_i$, then $p_c(W) = \prod (W - a_i)$, with $a_i = c^1(u_i)$.

Proposition 2.3. *Let $c = c(e)$. There is a canonical isomorphism*

$$\mathrm{Gr}\, K_e \approx (\mathrm{Gr}\, K)_c = (\mathrm{Gr}\, K)[W]/(p_c(W))$$

such that w corresponds to $W \bmod (p_c(W))$. The homomorphism

$$f_G: \mathrm{Gr}(K_e) \to \mathrm{Gr}(K)$$

has the property that

$$f_G(w^j) = \begin{cases} 0 & \text{if } 0 \leqq j < r, \\ 1 & \text{if } j = r. \end{cases}$$

In other words f_G is the functional g_c discussed in Chapter I, §3.

Proof. Proposition 1.1 shows that w is a root of $p_c(W)$. The isomorphism is then a consequence of Theorem 1.2.

By Corollary 1.3 the K-linear map f_e maps $F^k K_e$ into $F^{k-r}K$, so

$$f_G(w^j) = 0 \qquad \text{if} \quad j < r.$$

For $j = r$,

$$\begin{aligned} f_G(w^r) &= \varepsilon(f((\ell - 1)^r)) \\ &= \sum_{i=0}^{r} (-1)^{r-i} \varepsilon(f(\ell^i)) \binom{r}{i} \\ &= \sum_{i=0}^{r} (-1)^{r-i} \binom{r+i}{i} \binom{r}{i} \qquad \text{by Lemma 1.1 of Chapter I.} \end{aligned}$$

But this expression is the coefficient of t^r in the expansion of $1/(1-t)$, so is equal to 1, as one sees from the expressions

$$(1-t)^r = \sum (-1)^{r-i} \binom{r}{i} t^{r-i} \qquad \text{and} \qquad \frac{1}{(1-t)^{r+1}} = \sum_{j=0}^{\infty} \binom{r+j}{j} t^i.$$

This concludes the proof.

Corollary 2.4. *Let K be a λ-ring functor. Let $f: X \to Y$ be a morphism for which there exists a positive element $e \in K(Y)$ such that $K(X)$ is isomorphic to $K(Y)_e$ as a $K(Y)$-algebra via f^K, and such that f_K corresponds to f_e. Then f is an elementary projection with respect to the Chern class functor* $(K, c, \operatorname{Gr} K)$.

From Chapter II, Theorem 2.2 it follows that Riemann–Roch holds for f with respect to $(K, \operatorname{ch}, \mathbf{Q}\operatorname{Gr} K)$, with multiplier

$$\tau_f = \operatorname{td}(\ell e^\vee).$$

III §3. Adams Operations and the Filtration

Let K be a λ-ring. We want to see the effect of the Adams operations ψ^j on the graded ring $\operatorname{Gr}(K)$. Our goal is Theorem 3.5, which will combine properties of Adams and Chern characters.

Recall Proposition 6.3 of Chapter I:

$$\mathrm{ch}^k(\psi^j x) = j^k \, \mathrm{ch}^k(x).$$

This tells us that if the Chern character is to give an isomorphism between $\mathbf{Q}K$ and $\mathbf{Q}\mathrm{Gr}\, K$, then, if $j \geqq 2$, the eigenspace of ψ^j corresponding to eigenvalue j^k should map isomorphically onto $\mathbf{Q}\mathrm{Gr}^k\, K$.

The γ-filtration F^n of K induces a filtration

$$\mathbf{Q}F^n = \mathbf{Q} \otimes_{\mathbf{Z}} F^n$$

of

$$\mathbf{Q}K = \mathbf{Q} \otimes_{\mathbf{Z}} K.$$

Proposition 3.1. *Let $j \geqq 1$. Let n be an integer $\geqq 0$. If $x \in F^n$ then*

$$\psi^j(x) \equiv j^n x \mod F^{n+1}.$$

Hence $\mathrm{Gr}^n\, K$ is an eigenspace for $\mathrm{Gr}\, \psi^j$ with eigenvalue j^n.

Conversely, let $j \geqq 2$, and let $x \in \mathbf{Q}K$. If

$$\psi^j(x) \equiv j^n x \mod \mathbf{Q}F^{n+1}$$

then $x \in \mathbf{Q}F^n$.

Proof. For $n = 0$ the first assertion is immediate. Using the addition formula for the γ's, one sees easily that it suffices to prove this first assertion for elements of the form $x = \gamma^n(e)$ where e is positive. Using the assumption that there exists a λ-ring extension K' of K which splits e and such that

$$F^n K' \cap K = F^n K,$$

we see that we may assume e split. Again using the addition formula for the γ's, it suffices to prove the first assertion for elements of the form

$$x = (u_1 - 1) \cdots (u_n - 1),$$

where $u_1, \ldots, u_n$ are line elements. But then

$$\psi^j(x) = \prod_{i=1}^{n} (u_i^j - 1) = \prod_{i=1}^{n} (u_i - 1) \prod_{i=1}^{n} (1 + u_i + \cdots + u_i^{j-1})$$

and $1 + u_i + \cdots + u_i^{j-1} \equiv j \bmod F^1$, so the first part of the proposition follows.

As to the second, suppose $\psi^j(x) \equiv j^n x \bmod \mathbf{Q}F^{n+1}$. Let m be the largest integer such that $x \in \mathbf{Q}F^m$, and suppose $m < n$. We have

$$\psi^j(x) \equiv j^n x \mod \mathbf{Q}F^{n+1} \qquad \text{and} \qquad \psi^j(x) \equiv j^m x \mod \mathbf{Q}F^{m+1}.$$

Hence

$$(j^n - j^m)x \in \mathbf{Q}F^{m+1}$$

which contradicts the definition of m. This concludes the proof of the proposition.

We now let

$$V = \mathbf{Q}K = \mathbf{Q} \otimes_{\mathbf{Z}} K,$$

so V is a vector space over $\mathbf{Q}$. For each $j \geqq 2$ and each integer $m \geqq 0$ we let:

$V_j(m)$ = eigenspace for the operator ψ^j with eigenvalue j^m.

Proposition 3.2. *Assume that $F^{d+1} = 0$ for some integer d. Then the space $V_j(m)$ is independent of j, and so can be denoted $V(m)$, and*

$$\mathbf{Q}K = \bigoplus_{m=0}^{d} V(m).$$

Proof. By Proposition 3.1 we have for any integer $k \geqq 2$ and $m \geqq 0$:

$$\prod_{n \neq m} (\psi^j - j^n)(\psi^k - k^m) = 0,$$

and in the product, we actually have a finite product since we can take $n \leqq d$. Hence $V_j(m) \subset V_k(m)$, so we have equality by symmetry.

Again by Proposition 3.1,

$$\prod_{n=0}^{d} (\psi^j - j^n) = 0 \qquad \text{on } K,$$

and hence the left-hand side is also the 0 operator on V. Therefore there is a decomposition of the identity

$$\text{id} = \sum_{n=0}^{d} \prod_{m \neq n} (\psi^j - j^m)/(j^n - j^m).$$

The image of the m-th projection is $V(m)$. This concludes the proof.

Remark. Since $\psi^1(x) = x$ for all x, that is ψ^1 is the identity, it follows that the eigenspace $V(m)$ is also an eigenspace for ψ^1 with eigenvalue 1.

The following corollary merely gives a convenient reformulation of Proposition 3.2.

Corollary 3.3. *For $m \geqq 0$ we have a direct sum decomposition*

$$\mathbf{Q}F^m = V(m) \oplus \mathbf{Q}F^{m+1}.$$

We shall use this decomposition to get an isomorphism

$$\mathrm{ch}: \mathbf{Q}K \to \mathbf{Q}\mathrm{Gr}\, K.$$

Assume that $F^iK = 0$ for i sufficiently large. Define a map

$$g: \mathbf{Q}\mathrm{Gr}\, K \to \mathbf{Q}K$$

by defining it separately on each component, and for $x \in \mathbf{Q}\mathrm{Gr}^m$, let

$$g(x) = \text{unique element } x \text{ in } V(m) \text{ such that } x \equiv g(x) \bmod \mathbf{Q}F^{m+1}.$$

The existence and uniqueness of $g(x)$ follows at once from the decomposition of Corollary 3.3. Since g is well defined, it follows easily that g is a ring homomorphism.

Proposition 3.4. *If $x = u - 1 \bmod F^2$ with $u \in \mathbf{L}$, then*

$$g(x) = \log(1 + (u - 1)) = \sum (-1)^{n-1} \frac{(u-1)^n}{n}.$$

Proof. It is immediate that the right-hand side $\bmod F^2$ is equal to $u - 1$ in $F^1/F^2 = \mathrm{Gr}^1$. Since ψ^j is a ring homomorphism, we can apply ψ^j term by term to get the eigenspace property for the expression on the right-hand side, as desired.

Theorem 3.5. *Assume that $F^iK = 0$ for i sufficiently large. Then the maps*

$$\mathrm{ch}: \mathbf{Q}K \to \mathbf{Q}\mathrm{Gr}\, K \quad \textit{and} \quad g: \mathbf{Q}\mathrm{Gr}\, K \to \mathbf{Q}K$$

are inverse ring isomorphisms. In fact, for each integer $m \geqq 0$, ch induces a $\mathbf{Q}$-vector space isomorphism

$$\mathrm{ch}: V(m) \xrightarrow{\approx} \mathbf{Q}\mathrm{Gr}^m\, K.$$

Proof. We may pass to an extension which splits a given element. In that case, it suffices to prove that the two maps are inverse to each other on line elements u and $u - 1 \bmod F^2$. In this case the assertion is obvious from the definitions of ch and g.

For $x \in \mathbf{Q}K$ write

$$\mathrm{ch}(x) = \sum_{m \geqq 0} \mathrm{ch}^m(x)$$

with $\mathrm{ch}^m(x) \in \mathbf{Q}\mathrm{Gr}^m K$. For example, $\mathrm{ch}^0(x) = \varepsilon(x)$.

Proposition 3.6. *If* $\mathrm{ch}^i(x) = 0$ *for all* $i < m$, *then*

$$\mathrm{ch}^m(x) = x \mod \mathbf{Q}F^{m+1}K.$$

Proof. Let $\bar{x}$ be the image of x in $\mathbf{Q}\mathrm{Gr}^m K$. Then $g(\bar{x}) - x \in \mathbf{Q}F^{m+1}$, and

$$\mathrm{ch}^i(g(\bar{x}) - x) = 0 \qquad \text{for} \quad i \leqq m,$$

so $\bar{x} = \mathrm{ch}^m g(\bar{x}) = \mathrm{ch}^m(x)$, as was to be proved.

In a geometric context, the condition that $F^iK = 0$ for i sufficiently large will be proved in Chapter V, Corollary 3.10.

III §4. An Equivalence Between Adams and Grothendieck Riemann–Roch Theorems

In this section we let K *be a* λ*-ring functor. We suppose that for each* X, *there exists an integer* d *such that* $F^iK(X) = 0$ *for* $i > d$. *Since we work with rational coefficients, we write* $K(X)$ *for* $\mathbf{Q}K(X)$ *and* $G(X)$ *for* $\mathbf{Q}\mathrm{Gr}\, K(X)$.

It will be convenient to introduce the characters

$$\varphi^j : G(X) \to G(X),$$

which are multiplication by j^k on the k-th graded piece G^kX. Each φ^j is a ring homomorphism, and $\varphi^i \circ \varphi^j = \varphi^{i+j}$. If $f : X \to Y$ is a morphism, then $f^G\varphi^j = \varphi^j f^G$, while if $f_G : G(X) \to G(Y)$ raises degrees by d, then

$$\varphi^j f_G(x) = f_G(j^d \varphi^j x)$$

for $x \in G(X)$. (This trivial formula may be regarded as a Riemann-Roch formula for f with respect to (G, φ^j, G), with multiplier j^d.) Proposition 6.3 of Chapter I reads

$$\varphi^j \operatorname{ch}(x) = \operatorname{ch} \psi^j(x)$$

for $x \in K(X)$. Similarly $\varphi^j \operatorname{td}(x) = \operatorname{td} \psi^j(x)$.

Theorem 4.1. *Fix $j \geqq 2$. Let $f: X \to Y$ be a morphism, let*

$$\tau \in 1 + G^+(X),$$

and let d be a fixed integer. Then the following are equivalent:

(1) *f_K has degree d, and Riemann-Roch holds for f with respect to $(K, \operatorname{ch}, G)$, with multiplier τ.*
(2) *Riemann-Roch holds for f with respect to (K, ψ^j, K), with multiplier $\theta \in K(X)$ defined by*

$$\operatorname{ch}(\theta) = j^d \tau^{-1} \varphi^j(\tau).$$

Proof. Note that ch is an isomorphism, so the equation in (2) defines θ. Similarly let $z \in K(X)$ be defined by

$$\operatorname{ch}(z) = \tau^{-1}.$$

We shall use Theorem 3.5 as a matter of course, without further explicit reference.

Step 1. Suppose (2) holds. Then $f_K(z \cdot V(m)) \subset V(m+d)$. where $V(m)$ denotes the eigenspace of ψ^j with eigenvalue j^m. To see this, if $x \in V(m)$, then

$$\begin{aligned}
\psi^j f_K(z \cdot x) &= f_K(\theta \cdot \psi^j(zx)) \qquad \text{by (2)} \\
&= f_K(\operatorname{ch}^{-1}(j^d \tau^{-1} \varphi^j(\tau)) \cdot \psi^j \operatorname{ch}^{-1}(\tau^{-1}) \cdot \psi^j(x)) \\
&= f_K(j^d z \cdot \psi^j \operatorname{ch}^{-1}(\tau) \cdot \psi^j \operatorname{ch}^{-1}(\tau^{-1}) \cdot \psi^j(x)) \\
&= f_K(j^d z \cdot j^m x) \\
&= j^{d+m} f_K(z \cdot x),
\end{aligned}$$

as required.

Step 2. $(2) \Rightarrow (1)$. By Step 1, since

$$F^k K(X) = \bigoplus_{m \geqq k} V(m),$$

it follows that $f_K(F^kK(X)) \subset F^{k+d}K(Y)$. To finish, we must verify that, for any $y \in K(X)$,

$$\text{ch } f_K(y) = f_G(\tau \cdot \text{ch}(y)).$$

Let $x = z^{-1} \cdot y$, with z as above. The required formula is equivalent to showing

$$\text{ch } f_K(z \cdot x) = f_G(\text{ch}(x)).$$

It suffices to verify this for $x \in V(m)$, since $K(X)$ is a sum of such spaces. Then $f_K(z \cdot x)$ is in $V(m + d)$ by Step 1. But then $\text{ch}(x) \in G^mX$ is represented by $x \bmod F^{m+1}K(X)$, and $\text{ch } f_K(z \cdot x)$ is represented by $f_K(z \cdot x) \bmod F^{m+d+1}K(Y)$ (cf. Proposition 3.6). And

$$x \equiv z \cdot x \mod F^{m+1}K(X)$$

since $\varepsilon(z) = 1$. Hence $f_K(x) \equiv f_K(z \cdot x) \bmod F^{m+d+1}K(Y)$, and $\text{ch } f_K(z \cdot x)$ is represented by $f_K(x)$. Since x represents $\text{ch}(x)$, $f_G(\text{ch}(x))$ is also represented by $f_K(x) \bmod F^{m+d+1}K(Y)$, which completes the proof that (2) $\Rightarrow$ (1).

Step 3. (1) $\Rightarrow$ (2). Since ch is an isomorphism, (2) is equivalent to showing that

$$\text{ch } \psi^j f_K(x) = \text{ch } f_K(\theta \cdot \psi^j x)$$

for all $x \in K(X)$. Now

$$\begin{aligned}
\text{ch } f_K(\theta \cdot \psi^j x) &= f_G(\tau \cdot \text{ch}(\theta \cdot \psi^j x)) \qquad \text{by (1)} \\
&= f_G(\tau \cdot j^d \tau^{-1} \varphi^j(\tau) \cdot \text{ch}(\psi^j x)) \\
&= f_G(j^d \varphi^j(\tau \cdot \text{ch}(x))) \\
&= \varphi^j f_G(\tau \cdot \text{ch}(x)) \\
&= \varphi^j \text{ ch}(f_K(x)) \qquad \text{by (1)} \\
&= \text{ch}(\psi^j f_K(x)),
\end{aligned}$$

as required. This concludes the proof of Theorem 4.1.

To apply this theorem we will need elements τ and θ related as in (2). Such are provided by the following lemma.

Lemma 4.2. *Let K be a λ-ring. Let $q \in K$, with $\varepsilon(q) = d \in \mathbf{Z}$. Then for any $j \geqq 2$, we have*

$$\text{ch}(\theta^j(q)) = j^d \tau^{-1} \varphi^j(\tau),$$

where $\tau = \text{td}(q^\vee)^{-1}$.

Proof. Since both sides are homomorphic in q, by splitting it suffices to verify the formula when q is a line element. In that case, let $a = c^1(q)$. Then

$$\mathrm{ch}(\theta^j(q)) = 1 + e^a + \cdots + e^{(j-1)a},$$

$$\mathrm{td}(q^\vee) = \frac{-a}{1 - \mathbf{e}^a},$$

$$\varphi^j\,\mathrm{td}(q^\vee) = \frac{-ja}{1 - \mathbf{e}^{ja}},$$

and the lemma follows immediately.

From Lemma 4.2 and Theorem 4.1, we obtain:

Theorem 4.3. *Let* $f: X \to Y$ *be a morphism, let* $q \in K(X)$, *and* $d = \varepsilon(q)$. *The following are equivalent.*

(1) f *has graded degree* d *and Riemann–Roch holds for* f *with respect to* (K, ch, G) *with multiplier* $\mathrm{td}(q^\vee)^{-1}$.
(2) *For some* $j \geqq 2$, *Riemann–Roch holds for* f *with respect to* (K, ψ^j, K) *with multiplier* $\theta^j(q)$.
(3) *Same as* (2), *for all* $j \geqq 1$.

Remark. We shall use Adams Riemann–Roch in Chapter V, §6 to show that certain morphisms have graded degree.

CHAPTER IV

Local Complete Intersections

We now switch from abstract algebra to algebraic geometry.

This chapter describes in detail the basic category with which we shall deal in the context of algebraic geometry, namely regular morphisms. By this we mean morphisms which can be factored into a local complete intersection imbedding, and the projection from a projective bundle. Of course, it must be proved that such morphisms form a category. We study the basic geometric objects associated with such morphisms, namely the normal and tangent sheaves. Such sheaves are related by exact sequences, which will be interpreted in K-theory in Chapter V.

It is also natural to consider blow ups as part of the theory of projective bundles, and we give a concrete realization of the deformation of a regular imbedding to the normal bundle satisfying the axioms of Chapter II, §1 concerning basic deformations.

In this chapter, we use Koszul complexes in connection with regular sequences and regular imbeddings. In the next chapter, we shall use Koszul complexes to calculate K-groups explicitly.

IV §1. Vector Bundles and Projective Bundles

We first recall the basic notion $\mathrm{Proj}(\mathscr{S})$, where $\mathscr{S}$ is a sheaf of graded $\mathscr{O}_X$-algebras on a scheme X (cf. [H], II). Assume

$$\mathscr{S} = \bigoplus_{i \geqq 0} \mathscr{S}^i, \qquad \mathscr{S}^0 = \mathscr{O}_X,$$

$\mathscr{S}^1$ is a coherent sheaf of $\mathscr{O}_X$-modules, and $\mathscr{S}$ is locally generated by $\mathscr{S}^1$ as an algebra over $\mathscr{O}_X$. Then

$$\mathbf{P} = \mathrm{Proj}(\mathscr{S}), \qquad p\colon \mathbf{P} \to X$$

is a scheme over X, equipped with a canonical invertible sheaf $\mathscr{O}_{\mathbf{P}}(1)$ on $\mathbf{P}$. Locally X is $\mathrm{Spec}(A)$, and $\mathscr{S}$ corresponds to a finitely generated graded A-algebra S. Taking independent variables $T_0,\ldots,T_r$ corresponding to generators for S^1, we have

$$S = A[T_0,\ldots,T_r]/I$$

with some homogeneous ideal I. In this case $\mathbf{P}$ is the subscheme of $\mathbf{P}^r_A$ defined by the ideal I, and $\mathcal{O}(1)$ is the restriction of the canonical invertible sheaf on $\mathbf{P}^r_A$. In general $\mathbf{P}$ can be patched together from such local descriptions.

A graded sheaf $\mathcal{M}$ of $\mathcal{S}$-modules determines a sheaf of $\mathcal{O}_{\mathbf{P}}$-modules on $\mathbf{P}$, denoted $\mathcal{M}^\sim$. For example

$$\mathcal{O}_{\mathbf{P}}(d) = \mathcal{S}(d)^\sim.$$

where $\mathcal{S}(d)$ is the translated module whose k-th graded piece is $\mathcal{S}^{k+d}$.

A surjection $\mathcal{S} \to \mathcal{S}'$ of graded $\mathcal{O}_X$-algebras determines a closed imbedding

$$i\colon \mathbf{P}' = \operatorname{Proj}(\mathcal{S}') \hookrightarrow \operatorname{Proj}(\mathcal{S}) = \mathbf{P},$$

with $i^*\mathcal{O}_{\mathbf{P}}(1) = \mathcal{O}_{\mathbf{P}'}(1)$, and $p \circ i = p'$.

By a **locally free sheaf** $\mathcal{E}$ on X we shall always mean that $\mathcal{E}$ has finite rank in addition to being locally free. For such $\mathcal{E}$, we let

$$\mathbf{P}\mathcal{E} = \mathbf{P}(\mathcal{E}) = \operatorname{Proj}(\operatorname{Sym}\mathcal{E}), \qquad p\colon \mathbf{P}(\mathcal{E}) \to X$$

be the **associated projective bundle**. The natural action

$$\operatorname{Sym}(\mathcal{E}) \otimes \mathcal{E} \to \operatorname{Sym}(\mathcal{E})(1)$$

corresponds to a surjection of $p^*\mathcal{E}$ onto $\mathcal{O}_{\mathbf{P}}(1)$. Letting $\mathcal{H}$ be the kernel, this gives the **universal**, or **tautological**, exact sequence

$$0 \to \mathcal{H} \to p^*\mathcal{E} \to \mathcal{O}_{\mathbf{P}}(1) \to 0$$

on $\mathbf{P}$. If $\operatorname{rank}(\mathcal{E}) = r + 1$, then $\mathcal{H}$ is locally free of rank r, and we call $\mathcal{H}$ the **universal hyperplane sheaf** on $\mathbf{P}(\mathcal{E})$. For another description of $\mathcal{H}$, see Proposition 3.13.

The above sequence is universal in the following sense. If $f\colon Z \to X$ is a morphism and $\mathcal{L}$ is an invertible sheaf on Z, and

$$f^*\mathcal{E} \to \mathcal{L}$$

is a surjection, then there is a unique morphism $g: Z \to \mathbf{P}(\mathscr{E})$ with $p \circ g = f$, and an isomorphism of $g^*\mathcal{O}_{\mathbf{P}}(1)$ with $\mathscr{L}$, so that the diagram

$$\begin{array}{ccc} g^*p^*\mathscr{E} & \longrightarrow & g^*\mathcal{O}_{\mathbf{P}}(1) \\ \downarrow\approx & & \downarrow\approx \\ f^*\mathscr{E} & \longrightarrow & \mathscr{L} \end{array}$$

commutes. In particular, any surjection of $\mathscr{E}$ onto an invertible sheaf $\mathscr{L}$ on X determines a section $X \to \mathbf{P}$ of p.

Given $\mathscr{E}$, the above considerations apply to the locally free sheaf $\mathscr{E} \oplus \mathcal{O}_X$. We shall now globalize to $\mathbf{P}(\mathscr{E} \oplus \mathcal{O}_X)$ the simple concept of a hyperplane and its complement in projective space. Let

$$\psi: \mathbf{P}(\mathscr{E} \oplus \mathcal{O}_X) \to X$$

be the corresponding projective bundle, and let

$$0 \to \mathscr{Q} \to \psi^*(\mathscr{E} \oplus \mathcal{O}_X) \to \mathcal{O}(1) \to 0$$

be the universal exact sequence on $\mathbf{P}(\mathscr{E} \oplus \mathcal{O}_X)$. We call $\mathscr{Q}$ the **universal hyperplane sheaf** on $\mathbf{P}(\mathscr{E} \oplus \mathcal{O}_X)$.

The projection $\mathscr{E} \oplus \mathcal{O}_X \to \mathcal{O}_X$ on the second factor determines a canonical section

$$f: X \to \mathbf{P}(\mathscr{E} \oplus \mathcal{O}_X)$$

of ψ, called the **zero section**. This imbedding f will be our main example of the axiomatic notion of **elementary imbedding** introduced in Chapter II. Since $\mathscr{E}$ is the kernel of the projection from $\mathscr{E} \oplus \mathcal{O}_X$ to $\mathcal{O}_X$, we have

$$f^*\mathscr{Q} = \mathscr{E}.$$

The other projection $\mathscr{E} \oplus \mathcal{O}_X \to \mathscr{E}$ determines a closed imbedding

$$i: \mathbf{P}(\mathscr{E}) \to \mathbf{P}(\mathscr{E} \oplus \mathcal{O}_X)$$

called the **hyperplane at infinity**.

The **vector bundle** associated with $\mathscr{E}$ is defined to be

$$\pi: \mathbf{V}(\mathscr{E}) \to X \qquad \text{where} \qquad \mathbf{V}(\mathscr{E}) = \operatorname{Spec}(\operatorname{Sym} \mathscr{E}).$$

The surjection $\operatorname{Sym}(\mathscr{E}) \to \mathcal{O}_X$ sending $\mathscr{E}$ to 0 determines the **zero section**

$$g: X \to \mathbf{V}(\mathscr{E})$$

of π.

It should be remarked that $\mathscr{E}$ is the sheaf of sections of the *dual* of the bundle $\mathbf{V}(\mathscr{E})$.

The natural open imbedding

$$j: \mathbf{V}(\mathscr{E}) \to \mathbf{P}(\mathscr{E} \oplus \mathscr{O}_X)$$

gives a decomposition of $\mathbf{P}(\mathscr{E} \oplus \mathscr{O}_X)$ *into a disjoint union*

$$\mathbf{P}(\mathscr{E} \oplus \mathscr{O}_X) = \mathbf{V}(\mathscr{E}) \cup \mathbf{P}(\mathscr{E}),$$

after we identify $\mathbf{V}(\mathscr{E})$ *and* $\mathbf{P}(\mathscr{E})$ *with their images under* j *and* i *respectively.*

Locally, we describe the natural imbedding j as follows. Suppose

$$X = \operatorname{Spec}(A), \ \mathscr{E} = E^{\sim},$$

where E is a free A-module. Then

$$\mathbf{P}(\mathscr{E} \oplus \mathscr{O}_X) = \operatorname{Proj}(\operatorname{Sym}(E \oplus A)) = \operatorname{Proj}(S[T]),$$

where $S = \operatorname{Sym}(E)$, and T is an indeterminate. Now $i(\mathbf{P}(\mathscr{E}))$ is the subscheme defined by $T = 0$, and the complement is one of the basic affine open sets covering $\mathbf{P}(\mathscr{E} \oplus \mathscr{O}_X)$, namely $\operatorname{Spec}(S[T]_{(T)})$, where $S[T]_{(T)}$ is the subalgebra of $S[T]_T$ consisting of quotients of degree 0. Since

$$S[T]_{(T)} \approx S = \operatorname{Sym}(E),$$

this proves the first assertion locally. The compatibility of the morphisms then follows from the definitions.

We may summarize this in the diagram

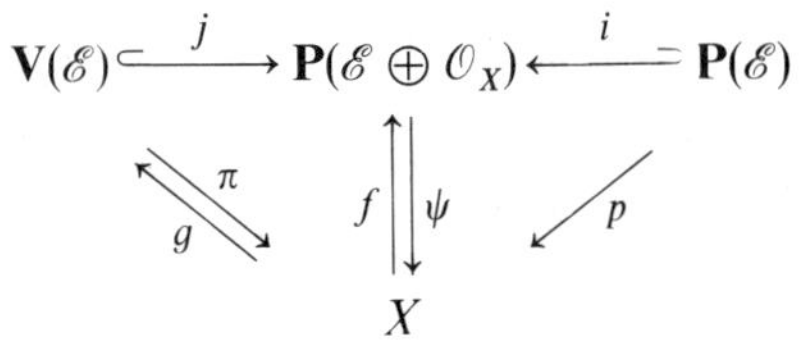

$$\begin{array}{ccccc} \mathbf{V}(\mathscr{E}) & \xhookrightarrow{\ j\ } & \mathbf{P}(\mathscr{E} \oplus \mathscr{O}_X) & \xleftarrow{\ i\ } & \mathbf{P}(\mathscr{E}) \\ & {}_{g}\nwarrow \searrow^{\pi} & f \uparrow \downarrow \psi & \swarrow_{p} & \\ & & X & & \end{array}$$

with the following commuting properties:

$$\psi \circ j = \pi, \qquad j \circ g = f, \qquad \psi \circ i = p.$$

We may call $\mathbf{P}(\mathscr{E} \oplus \mathscr{O}_X)$ the **projective completion** of $\mathbf{V}(\mathscr{E})$.

IV §2. The Koszul Complex and Regular Imbeddings

We start this section with general facts about Koszul complexes in commutative algebra. Such complexes give explicit resolutions, and lie at the base of what follows. We then translate this commutative algebra in the context of sheaves and give the applications to regular imbeddings.

Let A be a ring and E a finitely generated free module over A, of rank n. Let

$$E \xrightarrow{d_1} A$$

be a homomorphism. Let I be the image of d_1, so I is an ideal of A, and A/I is the cokernel of d_1. Then we may form the **Koszul complex**

$$0 \longrightarrow \Lambda^n E \xrightarrow{d_n} \Lambda^{n-1}E \longrightarrow \cdots \longrightarrow \Lambda^1 E \xrightarrow{d_1} A \longrightarrow 0,$$

where d_p is defined by the formula

$$d_p(t_1 \wedge \cdots \wedge t_p) = \sum_{j=1}^{p} (-1)^{j-1} d_1(t_j) t_1 \wedge \cdots \wedge \hat{t_j} \wedge \cdots \wedge t_p.$$

We shall determine conditions under which the Koszul complex is exact (except for the last term), and so gives a resolution of A/I. Note that we have not excluded the possibility that I is the unit ideal.

Suppose that I is generated by n elements, $I = (a_1,\ldots,a_n)$ such that if $e_1,\ldots,e_n$ is a basis of E then

$$a_i = d_1 e_i.$$

In terms of this basis, we let

$$K_p = \Lambda^p E = \text{free module with basis } \{e_{i_1} \wedge \cdots \wedge e_{i_p}\}, \qquad i_1 < \cdots < i_p.$$

Then the boundary

$$d_p \colon K_p \to K_{p-1}$$

is given by the formula

$$d_p(e_{i_1} \wedge \cdots \wedge e_{i_p}) = \sum_{j=1}^{p} (-1)^{j-1} a_{i_j} e_{i_1} \wedge \cdots \wedge \widehat{e_{i_j}} \wedge \cdots \wedge e_{i_p}.$$

Note that $K_0 = A$. In terms of the choice of a basis, or of the sequence $a_1,\ldots,a_n$, the Koszul complex is denoted by

$$K(a) \qquad \text{or} \qquad K(a_1,\ldots,a_n).$$

One may also construct $K(a)$ inductively:

$$K(a_1,\ldots,a_n) = K(a_1,\ldots,a_{n-1}) \otimes K(a_n).$$

We say that $(a) = (a_1,\ldots,a_n)$ is a **regular sequence** if $I \neq A$, if a_1 is not a divisor of 0 in A, and if the image of a_i in $A/(a_1,\ldots,a_{i-1})$ is not a divisor of zero.

By the **augmented Koszul complex**, we shall mean the complex

$$0 \to K_n \to K_{n-1} \to \cdots \to K_1 \to K_0 \to A/I \to 0.$$

where we stick A/I at the end.

Proposition 2.1.

(a) *If $a_1,\ldots,a_n$ is a regular sequence, then the augmented Koszul complex is exact, and so gives a resolution of A/I.*

(b) *If A is local and Noetherian, and $a_1,\ldots,a_n$ are in the maximal ideal of A, and the augmented Koszul complex is exact, then $a_1,\ldots,a_n$ is a regular sequence.*

(c) *If $I = (a_1,\ldots,a_n)$ is the unit ideal, then the Koszul complex $K(a)$ is exact.*

For a proof of (a), cf. [L], XVI, 10.4. Since that reference does not include a proof of (b) and (c), we do it here. We go back to the notation of [L], XVI, proof of Lemma 10.3. If C is a complex, and x an element of A, then there is an exact sequence for $p \geqq 0$:

$$\to H_{p+1}(C) \to H_{p+1}(C) \to H_{p+1}(C \otimes K(x)) \to H_p(C) \to H_p(C) \to H_p(C \otimes K(x)).$$

This exact sequence exists independently of any further assumptions on C. The map from $H_p(C)$ to $H_p(C)$ is multiplication by $(-1)^p x$, Then (a) is proved immediately from this sequence and induction.

We now prove (b). Assume that A is local Noetherian, and that $a_1,\ldots,a_n$ lie in the maximal ideal. Let

$$C = K(a_1,\ldots,a_{n-1}) \quad \text{and} \quad x = a_n.$$

We use the end of the exact sequence

$$H_1(C \otimes K(x)) \to H_0(C) \to H_0(C),$$

so the right arrow is multiplication by x. Since $H_1(C \otimes K(x))$ is assumed to be 0, it follows that multiplication by x is injective on $H_0(C)$,

which is $A/(a_1,\ldots,a_{n-1})$. Hence a_n is not divisor of 0 in that factor ring.
Furthermore, under the assumption that

$$H_{p+1}(C \otimes K(x)) = H_p(C \otimes K(x)) = 0$$

for $p \geqq 1$, the exact sequence implies that multiplication by x is an isomorphism on $H_p(C)$. Since x lies in the maximal ideal of a Noetherian ring, it follows that $H_p(C) = 0$ by Nakayama's lemma. The proof of (b) then follows by induction.

As to (c), we use the same type of technique. To prove that the Koszul complex is exact, it suffices to do so when we localize at each prime ideal of A, so we may assume that A is local. Under the assumption that I is the unit ideal, some element in the sequence is a unit. After reordering the sequence, say $x = a_n$ is a unit. In the long exact sequence, the map $H_p(C) \to H_p(C)$ is $(-1)^p a_n$, which is an isomorphism. Therefore $H_{p+1}(C \otimes K(x)) = 0$, thus proving that the Koszul complex is exact. This concludes the proof of Proposition 2.1.

The next theorem belongs to commutative algebra and will be applied to the geometric study of regular imbeddings and blow ups.

Let $A[X] = A[X_1,\ldots,X_n]$, and let Q be the ideal of $A[X]$ generated by

$$\{a_i X_j - a_j X_i\}, \qquad 1 \leqq i < j \leqq n.$$

Consider the canonical homomorphisms of graded A-algebras:

$$A[X]/Q \xrightarrow{\psi} \mathrm{Sym}_A(I) \xrightarrow{\varphi} \bigoplus_{m=0}^{\infty} I^m,$$

where

$$\psi(X_i) = a_i \in I = \mathrm{Sym}_A^1(I).$$

Theorem 2.2 (Micali). *If $a_1,\ldots,a_n$ is a regular sequence then ψ and φ are isomorphisms.*

Proof. By construction ψ and φ are surjective maps of graded algebras. It therefore suffices to show that if f is a homogeneous polynomial in $A[X]$ such that

$$f(a_1,\ldots,a_n) = 0,$$

then $f \in Q$. The proof will use the following lemma.

Lemma 2.3. *If* $f_1, \ldots, f_n \in A[X]$, *and* $\sum_{i=1}^{n} a_i f_i \in Q$, *then* $\sum_{i=1}^{n} X_i f_i \in Q$.

Proof. Let $A_k = A/(a_1, \ldots, a_k)$. By assumption there are $f_{i,j} \in A[X]$, $i < j$, with

$$\sum_{k=1}^{n} a_k f_k = \sum_{i<j} f_{i,j}(a_i X_j - a_j X_i),$$

or

$$(*) \qquad \sum_{k=1}^{n} a_k h_k = 0,$$

where

$$h_k = f_k - \sum_{k<j} f_{k,j} X_j + \sum_{i<k} f_{i,k} X_i.$$

We show by descending induction on k that there are $g_{i,k} \in A[X]$, $i < k$, with

$$(*)_k \qquad h_k + \sum_{j>k} a_j g_{k,j} = \sum_{i<k} a_i g_{i,k}$$

and

$$(*)'_k \qquad \sum_{i=1}^{k} a_i\left(h_i + \sum_{j>k} a_j g_{i,j}\right) = 0.$$

For $k = n$, since a_n is a non-zero-divisor on $A_{n-1}[X]$, from $(*)$ we find $g_{i,n} \in A[X]$, $i < n$, with

$$(*)_n \qquad h_n = \sum_{i<n} a_i g_{i,n}$$

and

$$(*)'_{n-1} \qquad \sum_{i=1}^{n-1} a_i(h_i + a_n g_{i,n}) = 0.$$

Inductively, if $(*)_{k+1}$ and $(*)'_k$ hold, since a_{k+1} is a non-zero-divisor in $A_k[X]$, there are $g_{i,k} \in A[X]$, $i < k$, with

$$(*)_k \qquad h_k + \sum_{j>k} a_j g_{k,j} = \sum_{i<k} a_i g_{i,k}$$

and

$$(*)'_{k-1} \qquad \sum_{i=1}^{k-1} a_i\left(h_i + \sum_{j \geqq k} a_j g_{i,j}\right) = 0.$$

This completes the inductive step.

Now using the definition of h_k, and then $(*)_k$:

$$\begin{aligned}\sum_{k=1}^{n} X_k f_k &= \sum_{k} X_k h_k + \sum_{k<j} f_{k,j} X_k X_j - \sum_{i<k} f_{i,k} X_i X_k \\ &= \sum_{k=1}^{n} X_k h_k \\ &= \sum_{i<k} g_{i,k} a_i X_k - \sum_{j>k} g_{k,j} a_j X_k \\ &= \sum_{i<j} g_{i,j}(a_i X_j - a_j X_i)\end{aligned}$$

which is in Q, as desired.

Now we conclude the proof of the theorem. Let f be homogeneous of degree m in $A[X]$ with $f(a_1, \ldots, a_n) = 0$. We may write

$$(**) \qquad f = \sum_{i=1}^{n} (X_i - a_i) f_i,$$

for some $f_i \in A[X]$, $\deg f_i \leqq m - 1$. Write

$$f_i = \sum_{j=0}^{m-1} b_{j,i},$$

where $b_{j,i}$ is homogeneous of degree j. Equating homogeneous terms of degree $0, 1, \ldots, m$ in $(**)$, we obtain

$$(**)_0 \qquad \sum_{i=1}^{n} a_i b_{0,i} = 0,$$

$$(**)_k \qquad \sum_{i=1}^{n} X_i b_{k-1,i} - \sum_{i=1}^{n} a_i b_{k,i} = 0, \qquad 0 < k < m,$$

$$(**)_m \qquad f = \sum_{i=1}^{n} X_i b_{m-1,i}.$$

Now by the lemma, from $(**)_0$ it follows that $\sum_{i=1}^{n} X_i b_{0,i} \in Q$. Applying the lemma and $(**)_k$ inductively, we have $\sum X_i b_{k,i} \in Q$ for $k = 1, 2, \ldots, m-1$. Then by $(**)_m$, $f \in Q$, as required.

Corollary 2.4. *If $a_1, \ldots, a_n$ is a regular sequence, then the canonical homomorphisms*

$$A/I[X_1, \ldots, X_n] \to \mathrm{Sym}_{A/I}(I/I^2) \to \bigoplus_{m=0}^{\infty} I^m/I^{m+1}$$

are isomorphisms.

Corollary 2.5. *The canonical homomorphism*

$$A[T_2, \ldots, T_n]/(a_1 T_2 - a_2, \ldots, a_1 T_n - a_n) \to A[a_2/a_1, \ldots, a_n/a_1]$$

which sends T_i to a_i/a_1, is an isomorphism.

The first corollary follows from the theorem by tensoring with A/I. The second follows from the theorem by inverting the image of X_1 and setting $T_i = X_i/X_1$.

For later use we insert the following lemma.

Lemma 2.6. *Let I be a proper ideal in a Noetherian local ring A which is generated by a regular sequence. Then any minimal set of generators for I forms a regular sequence.*

Proof. If $a_1, \ldots, a_n$ is a regular sequence generating I, any minimal sequence of generators of I must have the form $b_1, \ldots, b_n$, with

$$b_i = \sum_{i=1}^{n} \lambda_{ij} a_j,$$

$\lambda_{ij} \in A$, and $\Lambda = (\lambda_{ij})$ an invertible matrix. Then Λ determines an isomorphism of $K(a)$ with $K(b)$, which, by Proposition 2.1(b), concludes the proof.

The Koszul complex globalizes as follows. Let $\mathscr{E}$ be a locally free sheaf on X of rank n, and let

$$\mathscr{E} \xrightarrow{d_1} \mathscr{O}_X$$

be a homomorphism of $\mathscr{E}$ to the structure sheaf of X. We can form the **Koszul complex**

$$0 \longrightarrow \Lambda^n \mathscr{E} \xrightarrow{d_n} \Lambda^{n-1}\mathscr{E} \longrightarrow \cdots \longrightarrow \Lambda^1 \mathscr{E} \xrightarrow{d_1} \mathcal{O}_X \longrightarrow 0,$$

where d_p is defined by "contraction", namely

$$d_p(t_1 \wedge \cdots \wedge t_p) = \sum_{j=1}^{p} (-1)^{j-1} d_1(t_j) t_1 \wedge \cdots \wedge \hat{t_j} \wedge \cdots \wedge t_p.$$

If

$$\mathscr{E} \xrightarrow{d_1} \mathcal{O}_X \longrightarrow 0$$

is exact, then the Koszul complex is exact, because locally on X, it is just the same as the one constructed with a free module and we can apply Proposition 2.1(c). Hence we may say that the Koszul complex is the **Koszul resolution of $\mathcal{O}_X$ determined by** d_1.

On the other hand, let s be a section of $\mathscr{E}$. Then s determines a homomorphism

$$d_1 = s^\vee : \mathscr{E}^\vee \to \mathcal{O}_X.$$

The image of $s^\vee$ is a sheaf of ideals, which defines a closed subscheme of X denoted by $Z(s)$ and called the **zero scheme** of s. We then obtain the **Koszul complex** $K(s)$:

$$0 \longrightarrow \Lambda^n \mathscr{E}^\vee \longrightarrow \Lambda^{n-1}\mathscr{E}^\vee \longrightarrow \cdots \longrightarrow \Lambda^1 \mathscr{E}^\vee \xrightarrow{s^\vee} \mathcal{O}_X \longrightarrow 0$$

in which d_p is now defined by

$$d_p(t_1 \wedge \cdots \wedge t_p) = \sum_{j=1}^{p} (-1)^{j-1} t_j(s) t_1 \wedge \cdots \wedge \hat{t_j} \wedge \cdots \wedge t_p.$$

For $x \in X$ the stalk $\mathscr{E}_x$ is a free module over the local ring $\mathcal{O}_{x,X}$ of X at x. Taking a basis for $\mathscr{E}_x$, we can represent s by a sequence $a_1, \ldots, a_n$ of elements of $\mathcal{O}_{x,X}$. Then the stalk of $K(s)$ at x is isomorphic to $K(a)$. If $x \notin Z(s)$, then the Koszul complex is exact at x by Proposition 2.1(c); and by Proposition 2.1(a) and (b) for $x \in Z(s)$, it follows that the following conditions are equivalent, and define what we mean by a **regular section** s:

The Koszul complex $K(s)$ is a resolution of $\mathcal{O}_{Z(s)}$.

In the above local representation, the sequence $(a_1, \ldots, a_n)$ is regular at a point of $Z(s)$.

In this case, we call the following exact sequence the **Koszul resolution** of $\mathcal{O}_{Z(s)}$ **determined by** s:

$$0 \longrightarrow \Lambda^n \mathscr{E}^\vee \longrightarrow \Lambda^{n-1}\mathscr{E}^\vee \longrightarrow \cdots \longrightarrow \Lambda^1 \mathscr{E}^\vee \xrightarrow{s^\vee} \mathcal{O}_X \longrightarrow \mathcal{O}_{Z(s)} \longrightarrow 0.$$

Next we give an important example of a regular section.

Given a locally free sheaf $\mathscr{E}$ on X, consider the projective bundle $\psi: \mathbf{P}(\mathscr{E} \oplus \mathcal{O}_X) \to X$ with its universal exact sequence

$$0 \to \mathscr{Q} \to \psi^*\mathscr{E} \oplus \mathcal{O}_\mathbf{P} \to \mathcal{O}_\mathbf{P}(1) \to 0.$$

The dual of the first map gives a homomorphism from $\mathcal{O}_\mathbf{P} = \mathcal{O}_\mathbf{P}^\vee$ to $\mathscr{Q}^\vee$, which is a section s of $\mathscr{Q}^\vee$. We call s the **canonical section** of $\mathscr{Q}^\vee$.

Proposition 2.7. *The canonical section s of $\mathscr{Q}^\vee$ is regular, and its zero-scheme $Z(s)$ is $f(X)$, where f is the zero section imbedding of X in $\mathbf{P}(\mathscr{E} \oplus \mathcal{O}_X)$.*

Proof. The assertions are local on X, so we may assume $X = \operatorname{Spec}(A)$, and $\mathscr{E}$ is free with basis $T_1, \ldots, T_n$, so

$$\mathbf{P}(\mathscr{E} \oplus \mathcal{O}_X) = \operatorname{Proj}(A[T_0, \ldots, T_n]).$$

The zero-scheme $Z(s)$ is disjoint from the hyperplane $\mathbf{P}(\mathscr{E}) = Z(T_0)$ at infinity. On the complement

$$\mathbf{V}(\mathscr{E}) = \operatorname{Spec} A[T_1, \ldots, T_n],$$

$\pi = \psi \mid \mathbf{V}(\mathscr{E})$, $\mathscr{Q}$ restricts to $\pi^*\mathscr{E}$, and s is the tautological section of $\pi^*\mathscr{E}$, whose local equations are the regular sequence $T_1, \ldots, T_n$, which define the zero section of $\mathbf{V}(\mathscr{E})$, as required.

IV §3. Regular Imbeddings and Morphisms

In this section all schemes are Noetherian. Let $i: X \to Y$ be a *closed* imbedding, and let $\mathscr{I}$ be the ideal sheaf defining X in Y. The **conormal sheaf** $\mathscr{C}_{X/Y}$ to X in Y is the coherent sheaf of $\mathcal{O}_X$-modules defined by

$$\mathscr{C}_{X/Y} = \mathscr{I}/\mathscr{I}^2.$$

We say that i is a **regular imbedding** if every point of X has an affine neighborhood $\operatorname{Spec}(A)$ in Y such that the ideal of X in A is generated by a regular sequence.

Proposition 3.1. *Let $i: X \to Y$ be a closed imbedding. The following are equivalent:*

(i) *i is a regular imbedding.*
(ii) *Each point of X has a neighborhood U in Y such that there is a regular section of a locally free sheaf on U whose zero-scheme is $X \cap U$.*
(iii) *For each $x \in X$ the ideal $\mathscr{I}_x$ of X in $\mathcal{O}_{x,Y}$ is generated by a regular sequence.*
(iv) *For each $x \in X$ the ideal $\mathscr{I}_x \hat{\mathcal{O}}_{x,Y}$ in the completion $\hat{\mathcal{O}}_{x,Y}$ is generated by a regular sequence.*

Proof. The implications (i) $\Leftrightarrow$ (ii) $\Rightarrow$ (iii) are immediate from the definitions. For (iii) $\Rightarrow$ (i), choose an affine neighborhood $U = \operatorname{Spec}(A)$ of x such that there are elements $a_1, \ldots a_n$ in the ideal I of X in A which give a regular sequence of generators for $\mathscr{I}_x$. Shrinking U, one may assume $a_1, \ldots, a_n$ generate I. Consider the Koszul complex

$$0 \to K_n(a) \to \cdots \to K_0(a) \to A/I \to 0.$$

Since this complex is exact at x, it is exact in a neighborhood of x, for example since the support of the homology is closed and does not contain x.

The equivalence (iii) $\Leftrightarrow$ (iv) follows from the fact that, for $\mathcal{O} = \mathcal{O}_{x,Y}$, $\hat{\mathcal{O}}$ is flat over $\mathcal{O}$. Therefore if $a_1, \ldots, a_n$ is a minimal set of generators for $\mathscr{I}_x$, $K(a)$ is a resolution of $\mathcal{O}/\mathscr{I}_x\mathcal{O}$ if and only if

$$K(a) \otimes_{\mathcal{O}} \hat{\mathcal{O}} \to \hat{\mathcal{O}}/\mathscr{I}_x\hat{\mathcal{O}}$$

is a resolution. The proof concludes by Lemma 2.6, noting that a minimal set of generators for $\mathscr{I}_x$ is also a minimal set of generators for $\mathscr{I}_x\hat{\mathcal{O}}$.

Proposition 3.2.

(a) *If $i: X \to Y$ is a regular imbedding, then the conormal sheaf $\mathscr{C}_{X/Y}$ is locally free.*
(b) *If X is the zero scheme of a regular section of a locally free sheaf $\mathscr{E}$ on Y, then*

$$\mathscr{C}_{X/Y} \cong i^*\mathscr{E}^\vee.$$

Proof. (a) follows from Corollary 2.4 which implies that I/I^2 is free over A/I. For (b), consider the Koszul complex

$$\cdots \longrightarrow \Lambda^2\mathscr{E}^\vee \xrightarrow{d_2} \mathscr{E}^\vee \longrightarrow \mathscr{I} \longrightarrow 0.$$

Since the image of d_2 is contained in $\mathcal{I}\mathcal{E}^\vee$, tensoring by $\mathcal{O}_Y/\mathcal{I}$ gives the required isomorphism

$$i^*\mathcal{E}^\vee = \mathcal{E}^\vee \otimes \mathcal{O}_Y/\mathcal{I} \xrightarrow{\approx} \mathcal{I} \otimes \mathcal{O}_Y/\mathcal{I} = \mathcal{I}/\mathcal{I}^2.$$

Corollary 3.3. *If $\mathcal{E}$ is a locally free sheaf on a scheme X, then the zero section*

$$f: X \to \mathbf{P}(\mathcal{E} \oplus \mathcal{O}_X)$$

is a regular imbedding, with conormal sheaf $\mathcal{E}$.

Proof. This follows from Proposition 2.7, and the fact that $f^*\mathcal{Q} = \mathcal{E}$.

Proposition 3.4. *If $i: X \to Y$ and $j: Y \to Z$ are regular imbeddings, then $j \circ i: X \to Z$ is a regular imbedding, and there is an exact sequence*

$$0 \to i^*\mathcal{C}_{Y/Z} \to \mathcal{C}_{X/Z} \to \mathcal{C}_{X/Y} \to 0.$$

Proof. If $a_1, \ldots, a_m$ is a regular sequence generating an ideal I in a ring A, and $b_1, \ldots, b_n$ are elements in A whose images in A/I form a regular sequence, it follows immediately from the definition that $a_1, \ldots, a_m, b_1, \ldots, b_n$ is a regular sequence. This proves that the composite of regular imbeddings is regular. For any closed imbeddings $X \subset Y \subset Z$ one has an exact sequence of sheaves

$$\mathcal{C}_{Y/Z} \otimes_{\mathcal{O}_Y} \mathcal{O}_Z \to \mathcal{C}_{X/Z} \to \mathcal{C}_{X/Y} \to 0$$

on X. With regular sequences locally generating the ideals as above, one checks easily that this sequence is also exact on the left.

Remark. Given closed imbeddings $i: X \to Y$, $j: Y \to Z$, there are also partial converses to this proposition:

(i) *If $j \circ i$ and j are regular, and*

$$0 \to i^*\mathcal{C}_{Y/Z} \to \mathcal{C}_{X/Z} \to \mathcal{C}_{X/Y} \to 0$$

is exact and locally split, then i is regular.

(ii) *If $j \circ i$ and i are regular, then there is a neighborhood U of X in Z such that the imbedding of $Y \cap U$ in U is regular.*

For proofs the reader may consult [EGA], IV.19.1, or [SGA 6], VII.1.

Proposition 3.5. *Let $i\colon X \to Y$ be a closed imbedding, and let $f\colon Y' \to Y$ be a flat morphism. Form the fibre square:*

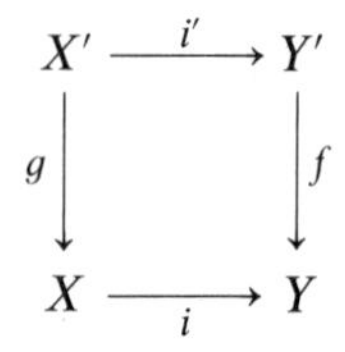

(a) *If i is a regular imbedding, then i' is a regular imbedding, and*

$$\mathscr{C}_{X'/Y'} = g^*\mathscr{C}_{X/Y}.$$

(b) *If f is surjective, and i' is a regular imbedding, then i is a regular imbedding.*

Proof. Recall that f is flat if, for all $y' \in Y'$, letting $y = f(y')$, $\mathcal{O} = \mathcal{O}_{y,Y}$, and $\mathcal{O}' = \mathcal{O}'_{y',Y'}$, $\mathcal{O}'$ is a flat $\mathcal{O}$-module. This means that if $\mathscr{K}.$ is an exact complex of $\mathcal{O}$-modules, then $\mathscr{K}. \otimes_{\mathcal{O}} \mathcal{O}'$ is also exact; in fact, since $\mathcal{O} \to \mathcal{O}'$ is a local homomorphism, the converse is also true, i.e. $\mathcal{O}'$ is faithfully flat over $\mathcal{O}$. Applying this when $\mathscr{K}.$ is a Koszul complex yields the proposition.

Let X and Y be schemes which are regularly imbedded in a scheme Z. We say that X and Y **meet regularly** if at each $x \in X \cap Y$, whenever $a_1, \ldots, a_m$ (resp. $b_1, \ldots, b_n$) is a regular sequence defining X (resp. Y) in Z near x, then

$$a_1, \ldots, a_m, b_1, \ldots, b_n$$

is a regular sequence defining $X \cap Y$ in Z near x. Equivalently, if s and t are regular sections of locally free sheaves $\mathscr{E}$ and $\mathscr{F}$ whose zero-schemes are X and Y near x, then $s \oplus t$ is a regular section of $\mathscr{E} \oplus \mathscr{F}$ whose zero-scheme is $X \cap Y$.

Proposition 3.6. *If X and Y meet regularly in Z, then the inclusions i, j, k of $X \cap Y$ in X, Y, Z are regular imbeddings, and*

$$\mathscr{C}_{X \cap Y/Z} \cong \mathscr{C}_{X \cap Y/X} \oplus \mathscr{C}_{X \cap Y/Y} \cong i^*\mathscr{C}_{Y/Z} \oplus j^*\mathscr{C}_{X/Z}.$$

Proof. This is essentially the same as the proof of Proposition 3.4, so will be omitted.

Recall that a morphism $f: X \to Y$ is **étale** if it is flat, and for all $x \in X$, $y = f(x)$, the induced homomorphism

$$\hat{\mathcal{O}}_{x,X} \to \hat{\mathcal{O}}_{y,Y}$$

of completions is an isomorphism. A morphism $f: X \to Y$ is **smooth** if, for each $x \in X$ there are neighborhoods U of x, V of $f(x)$, with $f(U) \subset V$, so that the restriction of f to U factors:

$$U \xrightarrow{g} \mathbf{A}_V^n \xrightarrow{p} V$$

with g étale and p the projection of a trivial vector bundle. We refer to [AK], VII for a readable account of basic properties of smooth morphisms.

A simple example of a smooth morphism is the projection morphism for a projective bundle or vector bundle. Such a morphism is locally isomorphic to a projection $\mathbf{A}_V^r \to V$. In fact, these bundle projections are the only smooth morphisms that are necessary for our treatment of Riemann–Roch, but we include general statements for completeness.

Recall also that for a morphism $f: X \to Y$, the **cotangent sheaf** $\Omega_{X/Y}^1$ is the conormal sheaf to the diagonal imbedding of X in $X \times_Y X$.

Proposition 3.7.

(a) *If $f: X \to Y$ is smooth, then $\Omega_{X/Y}^1$ is a locally free sheaf on X.*

(b) *If $f: X \to Y$ and $g: Y \to Z$ are smooth, then $g \circ f: X \to Z$ is smooth, and there is an exact sequence*

$$0 \to f^*\Omega_{Y/Z}^1 \to \Omega_{X/Z}^1 \to \Omega_{X/Y}^1 \to 0.$$

(c) *If $f: X \to Y$ is smooth, and $g: Y' \to Y$ is any morphism, form the fibre square*

$$\begin{array}{ccc} X' & \xrightarrow{f'} & Y' \\ {\scriptstyle g'}\downarrow & & \downarrow{\scriptstyle g} \\ X & \xrightarrow[f]{} & Y \end{array}$$

Then f' is smooth, and $\Omega_{X'/Y'}^1 = g'^*\Omega_{X/Y}^1$.

Proof. We note only that the assertions are evident for the case of bundle projections, and refer to [AK] for the general case.

Lemma 3.8. *If $f: X \to Y$ is smooth, and $i: Y \to X$ is a section of f, i.e. $f \circ i = \mathrm{id}_Y$, then i is a regular imbedding, and*

$$\mathscr{C}_{Y/Z} \cong i^*\Omega^1_{X/Y}.$$

Proof. When f is a projection

$$\mathbf{A}^n_V = \mathrm{Spec}(A[T_1,\ldots,T_n]) \to \mathrm{Spec}(A) = V$$

a section i is determined by the choice of $a_1,\ldots,a_n \in A$, mapping T_i to a_i. Then the ideal of Y in X is generated by $T_1 - a_1,\ldots,T_n - a_n$, which form a regular sequence in $A[T_1,\ldots,T_n]$. One sees directly in this case that the canonical homomorphism from $\mathscr{C}_{Y/X}$ to $i^*\Omega^1_{x/Y}$ is an isomorphism.

When f is étale, a section i of f must be a local isomorphism, so the lemma holds in this case. The general case follows easily from these two cases, for locally f is a composite

$$X \xrightarrow{g} \mathbf{A}^n_Y \xrightarrow{p} Y$$

of an étale g and a projection p. Form the fibre square

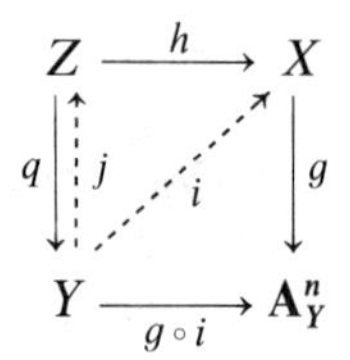

We have seen that $g \circ i$ is a regular imbedding. The section i determines a morphism $j: Y \to Z$ with $h \circ j = i$, $g \circ j = \mathrm{id}_Y$. Since q is étale, j is a local isomorphism. Since g is flat, h is a regular imbedding (Proposition 3.5). Therefore the composite $i = h \circ j$ is a regular imbedding. To check that the canonical map from $\mathscr{C}_{Y/X}$ to $i^*\Omega^1_{X/Y}$ is an isomorphism, one may localize and complete, so one is reduced to the first case.

Remark. In our applications, we shall deal with projective bundles, and only the first case will be relevant. This is the reason why we gave it first, with explicit coordinates. However, it is worth pointing out that once the section i has been proved to be regular as above, then the isomorphism

$$\mathscr{C}_{Y/X} \approx i^*\Omega^1_{X/Y}$$

can be seen directly as follows. We work locally, with $X = \mathrm{Spec}(A)$, $Y = \mathrm{Spec}(B)$, $A = B/I$ where I is the ideal of X in B and I is generated by a regular sequence. Then we have a map

$$I \to \Omega^1_{X/Y}/I\Omega^1_{X/Y} = \Omega^1_{X/Y} \otimes B/I$$

such that

$$b \mapsto db \mod I\Omega^1_{X/Y}.$$

The rule for the derivative of a product shows that I^2 is contained in the kernel. Given $b \in B$ there exists $a \in A$ such that $b \equiv a \bmod I$ and $db = d(b - a)$. Hence our map is surjective, and is a homomorphism of B/I-modules, that is A-modules. Both of these modules are free of the same rank, and hence the map is an isomorphism

$$I/I^2 \approx \Omega^1_{X/Y}/I\Omega^1_{X/Y}$$

as desired.

Proposition 3.9. *Consider a commutative triangle*

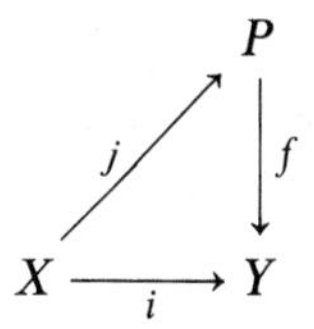

with f a smooth morphism, and i and j closed imbeddings. Then i is a regular imbedding if and only if j is a regular imbedding, in which case there is an exact sequence

$$0 \to \mathscr{C}_{X/Y} \to \mathscr{C}_{X/P} \to j^*\Omega^1_{P/Y} \to 0.$$

Proof. Form the fibre square

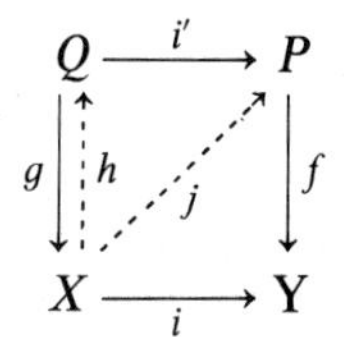

and let $h: X \to Q$ be the section of g determined by j. Then g is smooth (Proposition 3.7(c)), so h is a regular imbedding by the lemma, with

(i) $$\mathscr{C}_{X/Q} \cong h^*\Omega^1_{Q/X} = j^*\Omega^1_{P/Y}.$$

Suppose i is a regular imbedding. Then i' is regular by Proposition 3.5, with

$$\mathscr{C}_{Q/P} = g^*\mathscr{C}_{X/Y}.$$

Therefore

(ii) $$h^*\mathscr{C}_{Q/P} = h^*g^*\mathscr{C}_{X/Y} = \mathscr{C}_{X/Y}.$$

By Proposition 3.4 the composite $j = i' \circ h$ is then regular, with the sequence

$$0 \to h^*\mathscr{C}_{Q/P} \to \mathscr{C}_{X/P} \to \mathscr{C}_{X/Q} \to 0$$

exact. Combining this with (i) and (ii) gives the exact sequence asserted in the proposition.

Now assume that j is a regular imbedding. It remains to show that i must be regular. Factoring f locally as usual, it suffices to prove this when f is étale or a projection $\mathbf{A}^n_Y \to Y$. When f is étale, h is a local isomorphism, so i' is regular, and Proposition 3.5(b) implies i is regular. For a projection $\mathbf{A}^n_Y \to Y$, we may assume $Y = \operatorname{Spec}(A)$, with A local, so that h is the restriction of a section $Y \to \mathbf{A}^n_Y$ of f, given by $T_i \to a_i$ as in the proof of Lemma 3.8. If $b_1, \ldots, b_m$ is a minimal set of generators for the ideal of X in A, then

$$(b_1, \ldots, b_m, T_1 - a_1, \ldots, T_n - a_n)$$

is a minimal set of generators for the ideal of X in $A[T_1, \ldots, T_n]$. Since this sequence is regular in $A[T_1, \ldots, T_m]$, it follows that $b_1, \ldots, b_m$ is a regular sequence in A, as required.

Corollary 3.10. *Let $f: X \to Y$ be a morphism which admits two factorizations*

$$X \xrightarrow{\ i\ } P \xrightarrow{\ p\ } Y, \qquad X \xrightarrow{\ j\ } Q \xrightarrow{\ q\ } Y$$

with i and j closed imbeddings, and p and q smooth. Then i is regular if and only if j is regular.

Proof. Compare the two factorizations with the diagonal:

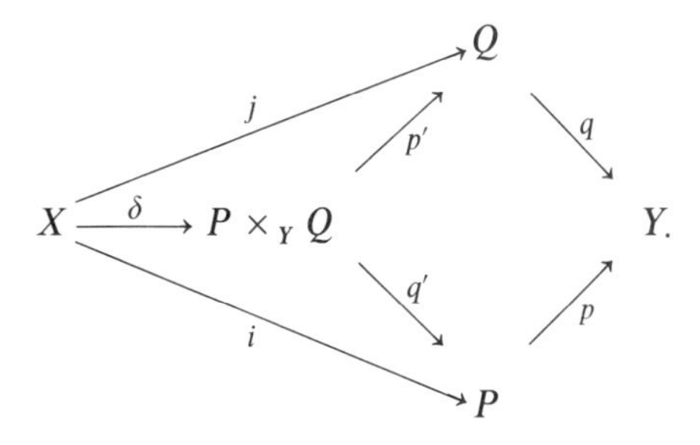

Since p' and q' are smooth, the proposition implies that the regularity of i or j is equivalent to the regularity of δ.

Proposition 3.11. *Consider a commutative triangle*

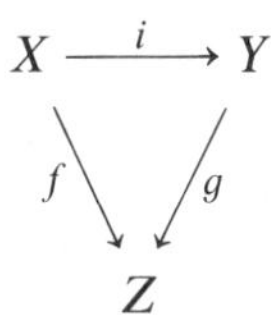

with f and g smooth, and i a closed imbedding. Then i is a regular imbedding, and there is an exact sequence

$$0 \to \mathscr{C}_{X/Y} \to i^*\Omega^1_{Y/Z} \to \Omega^1_{X/Z} \to 0.$$

Proof. Form the fibre square

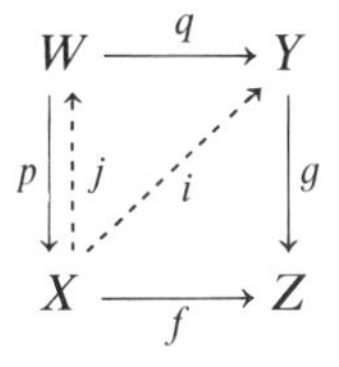

and let j be the section of p corresponding to i. Since p is smooth, j is regular (Lemma 3.8). Since q is smooth, it follows from the preceding proposition that i is regular, and

$$0 \to \mathscr{C}_{X/Y} \to \mathscr{C}_{X/W} \to j^*\Omega^1_{W/Y} \to 0$$

is exact. Since

$$\mathscr{C}_{X/W} = j^*\Omega^1_{W/X} = i^*\Omega^1_{Y/Z},$$

and

$$j^*\Omega^1_{W/Y} = j^*p^*\Omega^1_{X/Z} = \Omega^1_{X/Z},$$

the proposition follows.

Remark. Related to the last result is the general fact: *if $i\colon X \to Y$ is a closed imbedding of regular schemes, then i is a regular imbedding.*

This follows from the fact that if A is a regular local ring, and I is ideal in A such that A/I is regular, then I is generated by a regular sequence (cf. [Ma], 17.F).

We shall say that a morphism $f\colon X \to Y$ is a **regular morphism** if f factors into $p \circ i$:

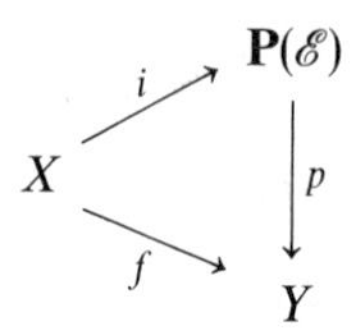

where $\mathscr{E}$ is a locally free sheaf on Y, p is the projection, and i is a regular imbedding. It is a consequence of Corollary 3.10 that if f is factored into any closed imbedding j followed by any smooth morphism q then j must be a regular imbedding, but we do not need this fact. Our regular morphisms are what are often called **projective local complete intersection morphisms.**

In case X is a scheme over a field k, X is called a **local complete intersection** if the structure morphism from X to Spec(k) is regular in the above sense. Note that such X need not be a regular scheme, although we shall see that local complete intersections do share several properties of non-singular varieties. Note also that fibres of a regular morphism need not be regular, or even local complete intersections.

In order to see that regular morphisms form a category, we need an additional assumption, which will be valid for all schemes X considered in the next chapter:

$(*)_X$ *Any coherent sheaf on X is the image of a locally free sheaf.*

Proposition 3.12. *If $f: X \to Y$ and $g: Y \to Z$ are regular morphisms, and $(*)_Q$ holds for all projective bundles Q over Z, then $g \circ f$ is also a regular morphism.*

Proof. Let $f: X \to Y$ be a regular morphism and $j: Y \to Q$ a closed imbedding. We claim first that granting $(*)_Q$, there is a locally free sheaf $\mathscr{E}$ on Q such that f factors into a regular imbedding

$$i: X \to \mathbf{P}(j^*\mathscr{E})$$

followed by the bundle projection from $\mathbf{P}(j^*\mathscr{E})$ to Y. To verify this, let $f = p_1 \circ i_1$ be any factorization of f into a regular imbedding i_1 of X in $P(\mathscr{E}_1)$ for some locally free sheaf $\mathscr{E}_1$ on Y, with p_1 the projection. Choose a surjection

$$\mathscr{E} \to j_*\mathscr{E}_1 \to 0$$

for some locally free sheaf $\mathscr{E}$ on Q. This surjection determines a closed imbedding of $\mathbf{P}(\mathscr{E}_1)$ in $\mathbf{P}(j^*\mathscr{E})$ which is regular by Proposition 3.11. By Proposition 3.4, the composite imbedding of X in $\mathbf{P}(j^*\mathscr{E})$ is regular, which proves the claim.

Now if f and g are regular morphisms, by the claim just proved we may find a commutative diagram (which one might call the **staircase decomposition** of $g \circ f$)

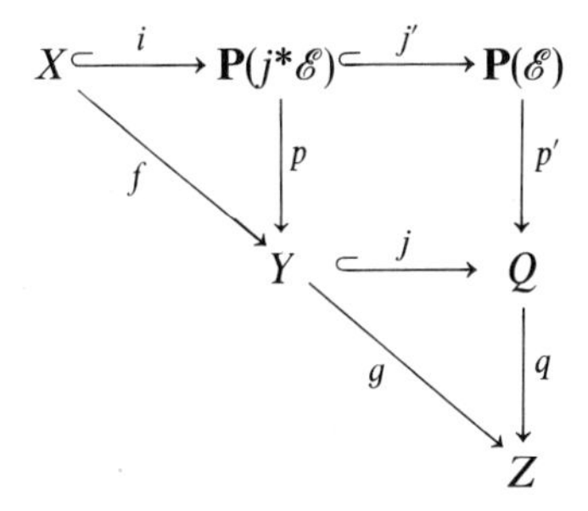

where the vertical maps are bundle projections and the horizontal maps are regular imbeddings. Then $g \circ f$ is the composite of $j' \circ i$ and $q \circ p'$. To conclude the proof, it suffices to show that $q \circ p'$ can be factored as a regular imbedding followed by a projective bundle projection.

In other words, we must show that if $\mathscr{F}$ is locally free on Z, $Y = \mathbf{P}\mathscr{F}$, $\mathscr{E}$ is locally free on Y, then the composite of the two projective bundle projections

$$f: X = \mathbf{P}\mathscr{E} \to Y \qquad \text{and} \qquad g: Y = \mathbf{P}\mathscr{F} \to Z$$

is a regular morphism. Indeed, for n sufficiently large, $g_*(\mathscr{E} \otimes \mathscr{O}_{\mathbf{P}\mathscr{F}}(n))$ is locally free on Z, and

$$g^*g_*(\mathscr{E} \otimes \mathscr{O}_{\mathbf{P}F}(n)) \to \mathscr{E} \otimes \mathscr{O}_{\mathbf{P}F}(n)$$

is surjective. This is a standard fact, but we shall reproduce a proof in Chapter V, §2, **R 4** and Proposition 2.2. This determines a closed imbedding

$$X \approx \mathbf{P}(\mathscr{E} \otimes \mathscr{O}_{\mathbf{P}\mathscr{F}}(n)) \hookrightarrow \mathbf{P}(g_*(\mathscr{E} \otimes \mathscr{O}_{\mathbf{P}\mathscr{F}}(n)),$$

as required.

Proposition 3.13. *Let $\mathscr{E}$ be a locally free sheaf on a scheme Y, $\mathbf{P} = \mathbf{P}(\mathscr{E})$ the associated projective bundle, $f\colon \mathbf{P} \to Y$ the projection, and*

$$0 \longrightarrow \mathscr{H} \xrightarrow{\;u\;} f^*\mathscr{E} \xrightarrow{\;v\;} \mathscr{O}_{\mathbf{P}}(1) \longrightarrow 0$$

the universal exact sequence. Then f is smooth, and

$$\Omega^1_{\mathbf{P}/Y} \cong \mathscr{H} \otimes \mathscr{O}_{\mathbf{P}}(-1), \qquad \text{so} \qquad \mathscr{H} \approx \Omega^1_{\mathbf{P}/Y}(1).$$

Proof. We have seen that f is smooth. Consider the diagonal

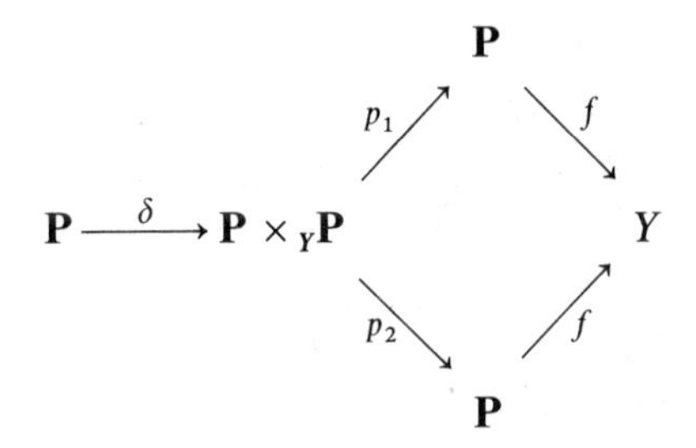

Note that $f \circ p_1 = f \circ p_2$. The composite

$$p_1^*\mathscr{H} \xrightarrow{\;p_1^*u\;} p_1^*f^*\mathscr{E} = p_2^*f^*\mathscr{E} \xrightarrow{\;p_2^*v\;} p_2^*\mathscr{O}(1)$$

is a section of the locally free sheaf

$$\mathscr{H}om(p_1^*\mathscr{H}, p_2^*\mathscr{O}(1)) = p_1^*\mathscr{H}^\vee \otimes p_2^*\mathscr{O}(1).$$

Looking locally, one sees that this section is a regular section, whose zero-scheme is precisely $\delta(\mathbf{P})$. It follows from Proposition 3.2 that

$$\mathscr{C}_{\mathbf{P}, \mathbf{P} \times_Y \mathbf{P}} = \delta^*(p_1^*\mathscr{H}^\vee \otimes p_2^*\mathscr{O}(1))^\vee$$

or equivalently

$$\Omega^1_{\mathbf{P}/Y} = \mathscr{H} \otimes \mathscr{O}(-1).$$

which proves the proposition.

Remark. As D. Laksov has pointed out, the same argument applies to general Grassmann bundles. If

$$\mathbf{G} = \operatorname{Grass}_d(\mathscr{E})$$

is the Grassmann bundle of rank d quotients of E, $f: \mathbf{G} \to Y$ the projection, and

$$0 \to \mathscr{S} \to f^*\mathscr{E} \to \mathscr{Q} \to 0$$

the universal exact sequence on $\mathbf{G}$, then f is smooth, and

$$\Omega^1_{\mathbf{G}/Y} = \mathscr{S} \otimes \mathscr{Q}^\vee.$$

Indeed, as above, $\mathbf{G}$ is the zero-scheme of a regular section of $\mathscr{H}om(p_1^*\mathscr{S}, p_2^*\mathscr{Q})$ on $\mathbf{G} \times_Y \mathbf{G}$.

A regular imbedding $i: X \to Y$ has **codimension** d at $x \in X$ if X is locally defined by a regular sequence of d elements near x; equivalently, $\mathscr{C}_{X/Y}$ is locally free of rank d is a neighborhood of x. Since the rank of $\mathscr{C}_{X/Y}$ is constant on connected components of X, the codimension is constant when X is connected.

A smooth morphism $f: X \to Y$ has **relative dimension** n at $x \in X$ if f factors locally near x into an étale morphism followed by a projection $\mathbf{A}^n_V \to V$; equivalently, $\Omega^1_{X/Y}$ is locally free of rank n in a neighborhood of x. When X is connected, the relative dimension is constant.

A regular morphism $f: X \to Y$ has **codimension** d if f factors into

$$X \xrightarrow{\ i\ } P \xrightarrow{\ p\ } Y,$$

where p is smooth (proper) morphism of some relative dimension r, and

i is a regular imbedding of codimension $d + r$. It follows from the proof of Corollary 3.10 that d is independent of the choice of factorization.

We mention another general fact, which has important applications for residues and duality but will not be used in this text.

Proposition 3.14. *Given a commutative triangle*

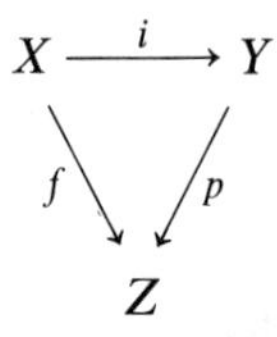

assume that p is smooth of relative dimension n, that i is a closed imbedding, with X locally defined in Y by n equations, and that f is finite. Then i is a regular imbedding, and f is flat, so $f_*\mathcal{O}_X$ is locally free on Z.

Proof. Let $x \in X$, and let A, B, C be the local rings of X, Y, Z at x, $i(x)$, $f(x)$. Let $b_1,\dots,b_n$ be a sequence of elements generating the ideal of X in B. Let $A_i = B/(b_1,\dots,b_i)$, so $A_0 = B$ and $A_n = A$. Let k be the residue field of C. Since p is smooth, $B \otimes_C k$ is a regular ring, in particular Cohen–Macaulay. Since

$$\dim(A \otimes_C k) = \dim(B \otimes_C k) - n$$

it follows that the images of $b_1,\dots,b_n$ in $B \otimes_C k$ form a regular sequence. Consider the exact sequences

$$A_i \xrightarrow{\varphi_i} A_i \longrightarrow A_{i+1} \longrightarrow 0,$$

where φ_i is multiplication by b_{i+1}. By the local criterion for flatness ([Mat], 20.E), from the injectivity of $\varphi_i \otimes_C k$ follows the injectivity of φ_i and the flatness of its cokernel A_{i+1}. This implies that $b_1,\dots,b_n$ is a regular sequence, and $A = A_n$ is flat over C.

Corollary 3.15. *In the situation of Proposition* 3.14, *for any base change $Z' \to Z$ the induced imbedding*

$$X \times_Z Z' \to Y \times_Z Z'$$

is a regular imbedding. In particular, a base change of a finite flat regular morphism is finite flat regular.

IV §4. Blowing Up

Let $i: X \to Y$ be a closed imbedding, and let $\mathscr{I}$ be the ideal sheaf of X in Y. The **blow up** of Y along X, denoted $\mathrm{Bl}_X Y$, or $\mathbf{B}$, is the projective scheme over Y constructed from the graded sheaf of $\mathcal{O}_Y$-algebra $\bigoplus \mathscr{I}^m$:

$$\mathrm{Bl}_X Y = \mathrm{Proj}\left(\bigoplus_{m \geqq 0} \mathscr{I}^m\right).$$

Let $\varphi: \mathrm{Bl}_X Y \to Y$ be the structure morphism, and let E be the **exceptional divisor**, i.e the inverse image of the scheme X:

$$E = \varphi^{-1}(X).$$

The fibre square

$$\begin{array}{ccc} E & \xrightarrow{j} & \mathrm{Bl}_X Y \\ \psi\downarrow & & \downarrow\varphi \\ X & \xrightarrow{i} & Y \end{array}$$

will be called the **blow up diagram** of the imbedding i. Let $\mathcal{O}(1)$ be the canonical invertible sheaf on $\mathrm{Bl}_X Y$ (§1).

Lemma 4.1.

(a) *As a scheme over X via ψ,*

$$E = \mathrm{Proj}\left(\bigoplus_{m \geqq 0} \mathscr{I}^m/\mathscr{I}^{m+1}\right).$$

The imbedding j of E in $\mathrm{Bl}_X Y$ is determined by the surjection

$$\bigoplus \mathscr{I}^m \to \bigoplus \mathscr{I}^m/\mathscr{I}^{m+1}$$

of graded algebras.

(b) *E is a Cartier divisor on $\mathrm{Bl}_X Y$, with ideal sheaf*

$$\mathcal{O}(-E) = \mathcal{O}(1).$$

(c) *If $f: Z \to Y$ is a morphism such that $f^{-1}(X)$ is a Cartier divisor on Z, then there is a unique $g: Z \to \mathrm{Bl}_X Y$ so that $f = \varphi \circ g$.*

Proof. For any $\varphi: \mathrm{Proj}(\mathscr{S}) \to Y$, and $X \subset Y$,

$$\varphi^{-1}(X) = \mathrm{Proj}(\mathscr{S} \otimes_Y \mathcal{O}_X).$$

Since $\mathscr{I}^m \otimes_Y \mathcal{O}_X = \mathscr{I}^m \otimes_Y \mathcal{O}_Y/\mathscr{I} = \mathscr{I}^m/\mathscr{I}^{m+1}$, this proves (a). For (b), the ideal sheaf of E in $\mathbf{B}$ is defined by the graded sheaf of ideals

$$\bigoplus_{m \geqq 0} \mathscr{I}^{m+1} = \mathscr{S}(1)$$

in $\mathscr{S}$, which defines $\mathcal{O}(1)$.

To prove (c), we may assume $Y = \operatorname{Spec}(A)$, X is defined by an ideal $I = (a_1, \ldots, a_n)$. Then we get a surjection

$$A[T_1, \ldots, T_n] \to \bigoplus I^m \to 0$$

by $T_i \mapsto a_i$, inducing a closed imbedding of $\mathrm{Bl}_X Y$ in $\mathbf{P}_Y^{n-1}$. By the universal property of projective bundles, f factors uniquely through $g: Z \to \mathbf{P}_Y^{n-1}$, by $T_i \mapsto f^*(a_i)$. One checks easily that this g factors through $\mathrm{Bl}_X Y$.

We shall require the case when i is a regular imbedding, so that we may take $a_1, \ldots, a_n$ to be a regular sequence. By Theorem 2.2, $\mathrm{Bl}_X Y$ is the subscheme of $\mathbf{P}_A^{n-1}$ defined by the equations.

$$a_i T_j - a_j T_i = 0, \qquad 1 \leqq i < j \leqq n.$$

The lemma is particularly evident in this case.

A consequence of (c) is the

Lemma 4.2. *If i imbeds X as a Cartier divisor on Y, then*

$$\varphi: \mathrm{Bl}_X Y \to Y$$

is an isomorphism.

Proposition 4.3.

(a) *If $i: X \to Y$ is a regular imbedding, with conormal sheaf $\mathscr{C}_{X/Y}$, then*

$$E = \mathbf{P}(\mathscr{C}_{X/Y}).$$

(b) *If X is the zero-scheme of a section of a locally free sheaf $\mathscr{E}$, there is a canonical imbedding of $\mathrm{Bl}_X Y$ into $\mathbf{P}(\mathscr{E}^\vee)$ over Y, which is a regular imbedding if i is a regular imbedding.*

(c) *Let $i: X \to Y$ be a regular imbedding. If the ideal sheaf is the image of a locally free sheaf (in particular, if condition $(*)_X$ at the end of §3 is satisfied) then the morphism $\mathrm{Bl}_X(Y) \to Y$ is a regular morphism.*

Proof. If i is regular, Corollary 2.4 yields

$$\mathrm{Sym}_{\mathcal{O}_X}(\mathcal{I}/\mathcal{I}^2) \approx \bigoplus_{m \geq 0} \mathcal{I}^m/\mathcal{I}^{m+1},$$

which proves (a). A section s of $\mathcal{E}$ determines a homomorphism $s^\vee$ from $\mathcal{E}^\vee$ to $\mathcal{O}_Y$ whose image is the ideal sheaf of the zero-scheme X. This induces a surjection of graded algebras

$$\mathrm{Sym}(\mathcal{E}^\vee) \to \bigoplus_{m \geq 0} \mathcal{I}^m \to 0$$

which determines the imbedding of $\mathrm{Bl}_X Y$ in $\mathbf{P}(\mathcal{E}^\vee)$.

If i is a regular imbedding, locally on Y we may write X as the zero-scheme of a regular section t of a locally free sheaf $\mathcal{F}$. Localizing further if necessary, we may assume $s^\vee$ factors into

$$\mathcal{E}^\vee \xrightarrow{\ u\ } \mathcal{F}^\vee \xrightarrow{\ t^\vee\ } \mathcal{O}_Y$$

for some surjective homomorphism u. This gives

$$\mathrm{Bl}_X Y \subset \mathbf{P}(\mathcal{F}^\vee) \subset \mathbf{P}(\mathcal{E}^\vee).$$

The first of these inclusions is a regular imbedding by Corollary 2.5, the second is clearly regular (Proposition 3.11), so the composite is regular (Proposition 3.4). This proves (b). Since a surjective homomorphism $\mathcal{E}^\vee \to \mathcal{I} \to 0$ is equivalent to a section of $\mathcal{E}$ whose zero scheme is X, (c) follows from (b).

Let E be closed subscheme of a scheme B. Assume E is a Cartier divisor on B. Let V be a closed subscheme of B which contains E as a subscheme:

$$E \subset V \subset B.$$

The **residual scheme** to E in V on B is defined to be the subscheme R of B whose ideal sheaf $\mathcal{I}(R)$ is related to the ideal sheaves $\mathcal{I}(E)$ and $\mathcal{I}(V)$ of E and V by the equation

$$\mathcal{I}(R) \cdot \mathcal{I}(E) = \mathcal{I}(V) \qquad \text{or} \qquad \mathcal{I}(R) = \mathcal{I}(V)\mathcal{I}(E)^{-1}.$$

Since $\mathcal{I}(E)$ is invertible, this determines $\mathcal{I}(R)$ uniquely. Note that R is a subscheme of V; local equations for R in B are obtained by dividing local equations for V by a local equation for E.

Let $X \hookrightarrow Y$ and $Y \hookrightarrow Z$ be closed imbeddings. Let $\mathbf{B} = \mathrm{Bl}_X Z$ be the blow up of Z along X, with exceptional divisor E, and let $\tilde{Y} = \mathrm{Bl}_X Y$ be the blow up of Y along X, with exceptional divisor $\tilde{X}$. We have the blow up diagrams:

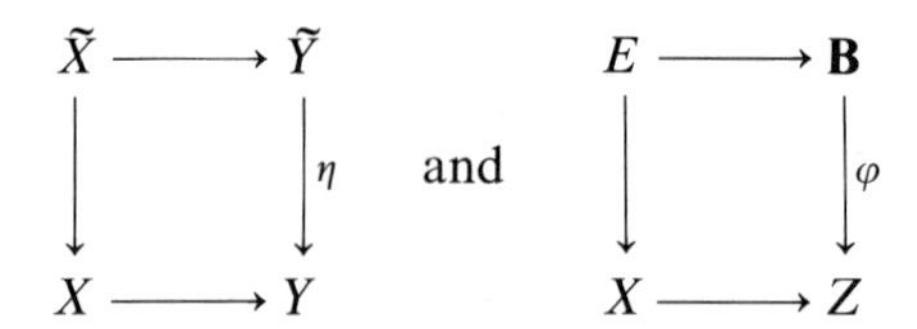

Lemma 4.4. *There is a unique closed imbedding of* $\tilde{Y} = \mathrm{Bl}_X Y$ *in* $\mathbf{B} = \mathrm{Bl}_X Z$ *such that the diagram*

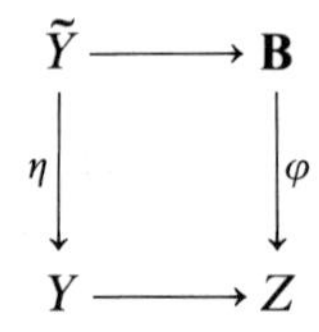

commutes, and $\tilde{X} = E \cap \tilde{Y}$.

Proof. If $\mathscr{I}$ is the ideal sheaf of X on Z, then $\mathscr{I}\mathcal{O}_Y$ is the ideal sheaf of X on Y, so there is a surjection of graded $\mathcal{O}_Y$-algebras

$$\bigoplus \mathscr{I}^m \otimes_Z \mathcal{O}_Y \to \bigoplus (\mathscr{I}\mathcal{O}_Y)^m \to 0.$$

This corresponds to an imbedding of $\tilde{Y}$ in $\varphi^{-1}(Y)$. The other assertions are special cases of Lemma 4.1(c).

One calls $\tilde{Y} = \mathrm{Bl}_X Y$ the **proper transform** of Y in $\mathbf{B} = \mathrm{Bl}_X Z$. It is a closed subscheme of the total transform $\varphi^{-1}(Y)$. Note that

$$E \subset \varphi^{-1}(Y) \subset \mathbf{B}.$$

Theorem 4.5. *If* $X \to Y$ *and* $Y \to Z$ *are regular imbeddings, then* $\tilde{Y}$ *is the residual scheme to* E *in* $\varphi^{-1}(Y)$ *on* $\mathbf{B}$. *In addition the imbedding of* $\tilde{Y}$ *in* $\mathbf{B}$ *is regular, with conormal sheaf*

$$\mathscr{C}_{\tilde{Y}/\mathbf{B}} = \eta^* \mathscr{C}_{Y/Z} \otimes \mathcal{O}_{\tilde{Y}}(\tilde{X}).$$

Proof. Assume that $Z = \mathrm{Spec}(A)$, and Y is defined by a regular sequence $a_1, \ldots, a_d$ in A, and X by a regular sequence $a_1, \ldots, a_n$, $n > d$. As we saw in the proof of Lemma 4.1, $\mathbf{B}$ is the subscheme of $\mathbf{P}_Z^{n-1}$ defined by the equations

$$a_i T_j = a_j T_i, \qquad 1 \leqq i < j \leqq n.$$

Similarly, $\tilde{Y}$ is the subscheme of $\mathbf{P}_Y^{n-1}$ defined by equations

$$\bar{a}_i T_j = \bar{a}_j T_i, \qquad d < i < j, \qquad \text{and} \qquad T_i = 0, \qquad 1 \leqq i \leqq d,$$

where $\bar{a}_i$ is the image of a_i in $A/(a_1,\ldots,a_d)$. Therefore the proper transform $\tilde{Y}$ is the subscheme of $\mathbf{B}$ defined by equations

$$T_1 = \cdots = T_d = 0, \qquad a_1 = \cdots = a_d = 0.$$

The total tranform $\varphi^{-1}(Y)$ is defined by equations

$$a_1 = \cdots = a_d = 0.$$

Let $U_k \subset B$ be the affine open set where $T_k \neq 0$, so

$$U_k = \operatorname{Spec}(A[t_1,\ldots,\hat{t}_k,\ldots,t_n]/(\{a_k t_i - a_i \mid i \neq k\})),$$

where $t_i = T_i/T_k$, by Corollary 2.5. On U_k the exceptional divisor E is defined by one equation $a_k = 0$. We must show that the ideals defining $\varphi^{-1}(Y)$, $\tilde{Y}$, and E, on U_k are related by the equation

$$I(\varphi^{-1}(Y)) = I(\tilde{Y}) \cdot I(E).$$

If $k \leqq d$, then Y is disjoint from U_k, and

$$I(\varphi^{-1}(Y)) = (a_1,\ldots,a_d) = (a_k) = I(E)$$

since $a_i = t_i a_k$. Similarly, if $k > d$,

$$\begin{aligned} I(\varphi^{-1}(Y)) &= (a_1,\ldots,a_d) = (t_1 a_k,\ldots,t_d a_k) \\ &= (t_1,\ldots,t_d)\cdot(a_k) = I(\tilde{Y})\cdot I(E). \end{aligned}$$

Since $t_1,\ldots,t_d$ form a regular sequence in the coordinate ring of U_k, $k > d$, it follows that the imbedding of $\tilde{Y}$ in $\mathbf{B}$ is regular.

It remains to verify the asserted relation between conormal sheaves. Starting with the residual relation

$$\mathscr{I}(Y)\mathcal{O}_{\mathbf{B}} = \mathscr{I}(\tilde{Y}) \cdot \mathscr{I}(E),$$

one deduces a surjection of sheaves on $\tilde{Y}$:

$$\eta^*(\mathscr{I}(Y)/\mathscr{I}(Y)^2) \to \mathscr{I}(\tilde{Y})/\mathscr{I}(\tilde{Y})^2 \otimes g^*\mathscr{I}(E),$$

where g is the inclusion of $\tilde{Y}$ in $\mathbf{B}$. Since both sides are locally free of the same rank, this surjection is an isomorphism. Since $E \cap \tilde{Y} = \tilde{X}$,

$$\eta^*\mathscr{C}_{X/Y} \approx \mathscr{C}_{\tilde{Y}/\mathbf{B}} \otimes \mathscr{O}_{\tilde{Y}}(-\tilde{X}),$$

as required.

IV §5. Deformation to the Normal Bundle

Let $f: X \to Y$ be a regular imbedding of codimension d, with conormal sheaf

$$\mathscr{C} = \mathscr{C}_{X/Y}.$$

Let

$$f': X \to Y' = \mathbf{P}(\mathscr{C}_{X/Y} \oplus \mathscr{O}_X) \tag{5.1}$$

be the zero section (see §1). We shall describe a deformation of f to f'. This "linearization" of f will be a concrete realization of the basic deformation considered in Chapter II, §1; we will construct a diagram

$$\begin{array}{ccccc} & & Y' & & \\ & \overset{f'}{\nearrow} & & \overset{g'}{\searrow} & \\ X & & & & M \\ & \underset{f}{\searrow} & & \underset{\pi}{\overset{g}{\nearrow\!\!\swarrow}} & \\ & & Y & & \end{array} \tag{5.2}$$

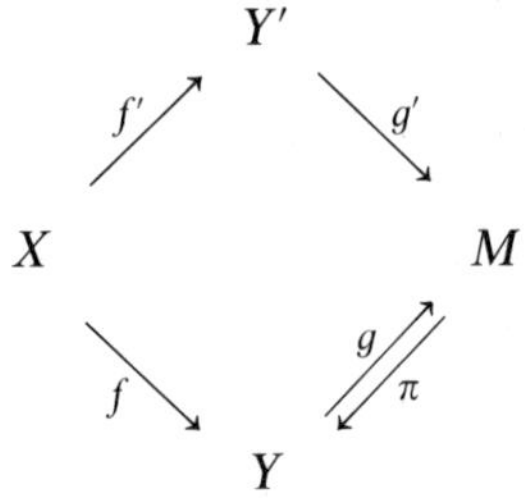

with $\pi \circ g = \mathrm{id}_Y$ and $\pi \circ g' \circ f' = f$.

First consider the projective line over Y:

$$p: \mathbf{P}^1_Y = \operatorname{Proj}(\mathscr{O}_Y[T_0, T_1]) \to Y.$$

There are two canonical sections of p:

$$s_0: Y \to \mathbf{P}^1_Y \qquad \text{and} \qquad s_\infty: Y \to \mathbf{P}^1_Y, \tag{5.3}$$

determined by $T_0 \mapsto 1$, $T_1 \mapsto 0$ and $T_0 \mapsto 0$, $T_1 \mapsto 1$ respectively. These sections each imbed Y as a Cartier divisor on $\mathbf{P}^1_Y$, with trivial conormal sheaf (Corollary 3.3). Let $Y(0) = s_0(Y)$, $Y(\infty) = s_\infty(Y)$, and

$$X(0) = s_0(f(X)), \qquad X(\infty) = s_\infty(f(X)).$$

Define M to be the blow up of $\mathbf{P}_Y^1$ along $X(\infty)$, that is

(5.4) $$M = \mathrm{Bl}_{X(\infty)}(\mathbf{P}_Y^1).$$

and let $\varphi: M \to \mathbf{P}_Y^1$ be the canonical morphism. Since

$$\mathbf{P}_X^1 \cap s_\infty(Y) = X(\infty),$$

it follows from Proposition 3.6 that the conormal sheaf to $X = X(\infty)$ in $\mathbf{P}_Y^1$ is given by

$$\mathscr{C}_{X/\mathbf{P}_Y^1} = \mathscr{C}_{X/Y} \oplus \mathscr{O}_X.$$

Therefore $Y' = \mathbf{P}(\mathscr{C} \oplus \mathscr{O}_X)$ is the exceptional divisor for the blow up $M \to \mathbf{P}_Y^1$, which determines an imbedding

(5.5) $$g': Y' \to M$$

making Y' a Cartier divisor on M. We get the blow up diagram:

$$\begin{array}{ccc} \mathbf{P}(\mathscr{C}_{X/Y} \oplus \mathscr{O}_X) & \xrightarrow{g'} & M \\ \psi \downarrow & & \downarrow \varphi \\ X & \xrightarrow[s_\infty \circ f]{} & \mathbf{P}_Y^1 \end{array}$$

We define $\pi: M \to Y$ to be the composite

(5.6) $$M \xrightarrow{\varphi} \mathbf{P}_Y^1 \xrightarrow{p} Y.$$

From the definition, $\pi \circ g' \circ f' = f$. Since φ is an isomorphism over the complement of $s_\infty(X)$, and $s_0(Y)$ is disjoint from $s_\infty(Y)$, the section s_0 determines a section

(5.7) $$g: Y \to M$$

of π, which makes Y a Cartier divisor on M. This completes the construction of the basic diagram.

From the composite

$$X \xrightarrow{f} Y \xrightarrow{s_\infty} \mathbf{P}_Y^1$$

we obtain a closed imbedding of $\tilde{Y} = \mathrm{Bl}_X Y$ in $\mathrm{Bl}_X(\mathbf{P}_Y^1)$, i.e.

(5.8) $$h: \tilde{Y} = \mathrm{Bl}_X Y \to M.$$

By Theorem 4.5, h is a regular imbedding of codimension 1, i.e. $\tilde{Y}$ is a Cartier divisor on M; and $\tilde{Y}$ is the residual scheme to Y' in $\varphi^{-1}(Y(\infty))$, i.e.

$$\mathscr{I}(\varphi^{-1}(Y(\infty))) = \mathscr{I}(\tilde{Y}) \cdot \mathscr{I}(Y').$$

Since these are ideal sheaves of Cartier divisors, the equality of ideal sheaves implies the equality

$$\varphi^*(Y(\infty)) = \tilde{Y} + Y' \tag{5.9}$$

of Cartier divisors on M. Since

$$\varphi^*(Y(0)) = Y, \tag{5.10}$$

imbedded in M by g, and $\mathcal{O}(Y(0)) \approx \mathcal{O}(Y(\infty)) \approx \varphi^*\mathcal{O}(1)$, we deduce:

$$\mathcal{O}(Y) \approx \mathcal{O}(\tilde{Y} + Y') = \mathcal{O}(\tilde{Y}) \otimes \mathcal{O}(Y'). \tag{5.11}$$

It also follows from Theorem 4.5 that the exceptional divisor $\mathbf{P}(\mathscr{C} \oplus \mathcal{O}_X)$ intersects $\mathrm{Bl}_X Y$ regularly in the scheme $\mathbf{P}(\mathscr{C})$:

$$Y' \cap \tilde{Y} = \mathbf{P}(\mathscr{C}). \tag{5.12}$$

Note that $\mathbf{P}(\mathscr{E})$ is imbedded in $\tilde{Y} = \mathrm{Bl}_X Y$ as the exceptional divisor, and in $Y' = \mathbf{P}(\mathscr{C} \oplus \mathcal{O}_X)$ as the hyperplane at infinity.

The imbedding of X in Y induces an imbedding of $\mathbf{P}^1_X$ in $\mathbf{P}^1_Y$. From the composite

$$X \xrightarrow{s_\infty} \mathbf{P}^1_X \longrightarrow \mathbf{P}^1_Y$$

we obtain an imbedding of $\mathrm{Bl}_X \mathbf{P}^1_X$ in $\mathrm{Bl}_X \mathbf{P}^1_Y$ (Lemma 4.4). Since X is a Cartier divisor on $\mathbf{P}^1_X$, $\mathrm{Bl}_X \mathbf{P}^1_X$ is just $\mathbf{P}^1_X$ (Lemma 4.2), so we have a closed imbedding

$$F : \mathbf{P}^1_X \to M \tag{5.13}$$

which is regular by Proposition 4.5.

We may summarize the results of this section in the **deformation diagram**

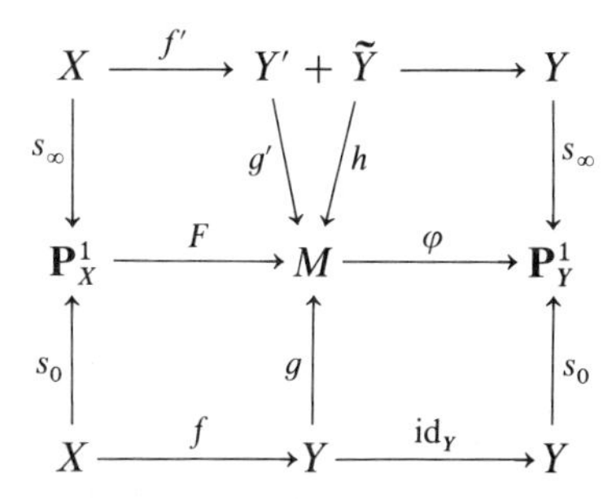

Each square in this diagram is a fibre square; the vertical maps are imbeddings of Cartier divisors. Note that $f'(X)$ is contained in Y' and is disjoint from $\tilde{Y}$.

Remark. If $f: X \to Y$ is a closed imbedding which is not necessarily regular, the same constructions yield a deformation diagram with exactly the same properties. The only difference is that $Y' = \mathbf{P}(\mathscr{C} \oplus \mathscr{O}_X)$ must be replaced by the projective completion of the normal *cone* to X in Y, i.e.

$$Y' = \operatorname{Proj}(\mathscr{S}[T]),$$

where $\mathscr{S} = \bigoplus_{m \geqq 0} \mathscr{I}^m/\mathscr{I}^{m+1}$, $\mathscr{I}$ the ideal sheaf of X in Y, and T is a variable of degree 1. In this case f' and F are closed imbeddings which may also not be regular. There are no essential changes in the proof, although Theorem 4.5 as stated does not apply in this situation. In fact, if $Y = \operatorname{Spec}(A)$, I is the ideal of X, and one identifies the complement of $Y(0)$ in $\mathbf{P}^1_Y$ with

$$\mathbf{A}^1_Y = \operatorname{Spec} A[T],$$

then the complement of $\varphi^{-1}(Y(0))$ in M is

$$\operatorname{Proj}\left(\bigoplus_{n=0}^{\infty} (I, T)^n\right),$$

from which it is easy to verify (5.9) and (5.12) (cf. [F 2], 5.1).

CHAPTER V

The K-Functor in Algebraic Geometry

In the previous chapter we described the geometry of regular morphisms. Here we describe one part of their homology, namely the K-functor on the category of locally free sheaves (which are always assumed to be of finite rank). An arbitrary locally free sheaf does not behave well under the direct image. Fortunately, the K-group generated by the locally free sheaves is also generated by a subfamily of sheaves which do behave well, and which we call regular sheaves. We use a cohomological characterization for these, due to Mumford.

These regular sheaves allow us first to determine $K(\mathbf{P}(\mathcal{E}))$ for a projective bundle as an algebra over K of the base. Then we use them to analyze the filtration. Finally they are used to prove certain properties of the push-forward maps on the K-groups. Among other things, we are in a position to prove that for regular morphisms f, the push-forward map f_K satisfies

$$(g \circ f)_K = g_K \circ f_K$$

as well as the projection formula. All the work has then been done to obtain the Grothendieck Riemann–Roch theorem in a few lines.

In this chapter, all schemes are assumed to be Noetherian, and connected (cf. Appendix). In particular, the rank of a locally free sheaf is constant.

Locally Free Resolutions

For the convenience of the reader, we shall now recall some facts about locally free resolutions. The proofs are obtained by replacing script letters for sheaves by latin letters for modules, in which case they are proved in [L], IV, §3, §8, and XVI, Theorem 3.8. The proofs for modules use a fact which we formulate here for sheaves:

A coherent sheaf is locally free if and only if it is flat.

Concerning resolutions, we have the following statements.

Let

$$0 \to \mathscr{E}' \to \mathscr{E} \to \mathscr{E}'' \to 0$$

be an exact sequence of coherent sheaves, and $\mathscr{E}$, $\mathscr{E}''$ locally free. Then $\mathscr{E}'$ is locally free.

In the reduction to the statement about modules, one merely has to look at the stalk at each point, which is a module over a local ring, and use the statement: *If $\mathscr{F}$ is a coherent sheaf, $\mathscr{F}_x$ the stalk at a point, and $\mathscr{F}_x$ is free, then $\mathscr{F}$ is free in a neighborhood of x.*

From the above property of short exact sequences, we conclude:

A long exact sequence

$$0 \to \mathscr{E}_n \to \mathscr{E}_{n-1} \to \cdots \to \mathscr{E}_0 \to 0$$

of locally free sheaves can always be decomposed into short exact sequences

$$0 \to \mathscr{K}_p \to \mathscr{E}_p \to \mathscr{K}_{p-1} \to 0,$$

where $\mathscr{K}_p$ and $\mathscr{K}_{p-1}$ are the kernel and cokernel respectively, and are locally free.

The above property of short exact sequences has a generalization to longer sequences:

Let $\mathscr{F}$ be a coherent sheaf on X which admits a resolution of length $\leqq n$ by locally free sheaves

$$0 \to \mathscr{E}_n \to \cdots \to \mathscr{E}_0 \to \mathscr{F} \to 0.$$

Let

$$0 \to \mathscr{Q} \to \mathscr{F}_{n-1} \to \cdots \to \mathscr{F}_0 \to \mathscr{F} \to 0$$

be a resolution where $\mathscr{F}_0,\ldots,\mathscr{F}_{n-1}$ are locally free. Then $\mathscr{Q}$ is locally free.

Basically, the proof for modules comes from "dimension shifting", whereby $\mathscr{F}$ has "dimension" $\leqq n$ implies $\mathscr{Q}$ has "dimension" 0.

In practice, the category of locally free sheaves also satisfies the property that given a coherent sheaf $\mathscr{F}$ on X, there exists a locally free sheaf $\mathscr{E}$ and a surjection

$$\mathscr{E} \to \mathscr{F} \to 0.$$

This is true for instance when there exists an ample sheaf on X, so when X is quasi-projective over an affine scheme. Since the category of locally free sheaves is closed under direct sums, it satisfies the three conditions under which one can do the basic K-theory, and is called a K-family in [L], IV, §3, where some general facts of K-theory are proved using only the above three conditions. In this chapter we shall see how they apply in the geometric context.

V §1. The λ-Ring $K(X)$

Let X be a scheme. We let:

$\mathfrak{V}_X$ = category of locally free sheaves on X;

$K(X)$ = Grothendieck group of $\mathfrak{V}_X$.

Thus $K(X)$ is the free abelian group on isomorphism classes $\{\mathscr{E}\}$ of locally free sheaves $\mathscr{E}$, modulo the subgroup generated by

$$\{\mathscr{E}\} - \{\mathscr{E}'\} - \{\mathscr{E}''\}$$

for each exact sequence

$$0 \to \mathscr{E}' \to \mathscr{E} \to \mathscr{E}'' \to 0 \tag{1.1}$$

of locally free sheaves on X. We shall write $[\mathscr{E}]$ for the class in $K(X)$ defined by $\mathscr{E}$. *Warning*: The map from isomorphism classes of locally free sheaves into $K(X)$ is not necessarily injective.

In Proposition 4.1 we shall see that $K(X)$ can also be described as the Grothendieck group of a category of certain coherent sheaves.

Tensor product induces a ring structure on $K(X)$ by the formula

$$[\mathscr{E}][\mathscr{F}] = [\mathscr{E} \otimes \mathscr{F}].$$

Exterior powers induce λ-operations:

$$\lambda^i[\mathscr{E}] = [\Lambda^i \mathscr{E}].$$

The rank function gives the augmentation ε:

$$\varepsilon([\mathscr{E}]) = \operatorname{rank}(\mathscr{E}).$$

The positive elements **E** are the classes $[\mathscr{E}]$ of locally free sheaves.

To see that $K(X)$ is a λ-ring, we must verify that for an exact sequence (1.1), we have

$$[\Lambda^k \mathscr{E}] = \sum_{i+j=k} [\Lambda^i \mathscr{E}' \otimes \Lambda^j \mathscr{E}''].$$

To see this, let $\mathscr{F}^i$ be the image of the canonical homomorphism

$$\Lambda^i \mathscr{E}' \otimes \Lambda^{k-i} \mathscr{E} \to \Lambda^k \mathscr{E}.$$

This gives a filtration $\Lambda^k \mathscr{E} = \mathscr{F}^0 \supset \mathscr{F}^1 \supset \cdots \supset \mathscr{F}^{k+1} = 0$ of $\Lambda^k \mathscr{E}$ by locally free subsheaves, with

$$\Lambda^i \mathscr{E}' \otimes \Lambda^{k-i} \mathscr{E}'' \cong \mathscr{F}^i / \mathscr{F}^{i+1}.$$

Note that the multiplicative group of line elements $\mathbf{L}$ is precisely the image of the natural map

$$\operatorname{Pic}(X) \to K(X),$$

where $\operatorname{Pic}(X)$ is the multiplicative group of isomorphism classes of invertible sheaves on X. In fact, we have an isomorphism

$$\operatorname{Pic}(X) \approx \mathbf{L}.$$

Indeed, the natural map is injective because the det map $\mathscr{E} \mapsto \Lambda^{\text{top}} \mathscr{E}$ as in the proof of Theorem 1.7 of Chapter III is now seen to induce a homomorphism of K into Pic(X) rather than of K into $\mathbf{L}$, and the composite map

$$\operatorname{Pic}(X) \longrightarrow K(X) \xrightarrow{\det} \operatorname{Pic}(X)$$

is the identity. Thus the isomorphism classes of invertible sheaves behave better in this respect than isomorphism classes of locally free sheaves, which do not map injectively in $K(X)$ because the Grothendieck relations identify different extensions. There is no room for extensions when dealing with invertible sheaves.

There is also an involution on $K(X)$, defined by

$$[\mathscr{E}]^\vee = [\mathscr{E}^\vee],$$

where $\mathscr{E}^\vee = \mathscr{H}om_{\mathscr{O}_X}(\mathscr{E}, \mathscr{O}_X)$ is the dual sheaf.

The association $X \mapsto K(X)$ is a contravariant functor. For any morphism $f\colon Y \to X$ we have the pull-back $f^*\mathscr{E}$ which is a locally free

$\mathcal{O}_Y$-module of the same rank as $\mathscr{E}$. This pull-back preserves a short exact sequence, and hence there is a unique additive homomorphism

$$f^K: K(X) \to K(Y)$$

such that $f^K[\mathscr{E}] = [f^*\mathscr{E}]$. Since f^* commutes with the tensor and alternating products up to isomorphism, we see that f^K is also a homomorphism of λ-rings with involution, and behaves functorially.

The λ-rings $K(X)$ satisfy all the properties stipulated in Chapter I. Only the (graded) splitting property and facts about the γ-filtration $F^n K(X)$ are not evident from the definitions; these will be verified in the next two sections.

Principally, we have to develop the covariant functorial properties of $X \mapsto K(X)$. In §2 we do what is necessary for the case of projections from a projective bundle, and in §4 we do what is necessary for regular embeddings. The combination of §2 and §4 in §5 will then show that our functor is a λ-ring functor on the category of regular morphisms under one other mild assumption that every coherent sheaf is the image of a locally free sheaf.

V §2. Sheaves on Projective Bundles

Throughout this section, we let $\mathscr{E}$ be a locally free sheaf of rank $r + 1$ on a scheme X. We let

$$\mathbf{P} = \mathbf{P}(\mathscr{E}) = \operatorname{Proj}(\operatorname{Sym} \mathscr{E}) \xrightarrow{\ f\ } X$$

be the associated projective bundle, and we let

$$\mathcal{O}(1) = \mathcal{O}_{\mathbf{P}}(1)$$

be the tautological invertible sheaf (Chapter IV, §1). *For any sheaf $\mathscr{F}$ of $\mathcal{O}_{\mathbf{P}}$-modules, and $n \in \mathbf{Z}$ we let*

$$\mathscr{F}(n) = \mathscr{F} \otimes_{\mathcal{O}_{\mathbf{P}}} \mathcal{O}(1)^{\otimes n},$$

where for $n < 0$, one defines $\mathcal{O}(1)^{\otimes n} = (\mathcal{O}(1)^{\vee})^{\otimes(-n)}$. We let $\mathfrak{V}_X$ be the category of locally free sheaves on X.

The main point of this section is to determine $K(\mathbf{P})$ as an algebra over $K(X)$, and to show that $K(\mathbf{P})$ is isomorphic to the extension $K(X)_e$ described in Chapter II, with $e = [\mathscr{E}]$. To determine the additive structure of $K(\mathbf{P})$ over $K(X)$ we follow Quillen's method, which is not only simpler for $K = K_0$, but also works for higher K groups, see [Q], §8, Theorem 3.1. The splitting property will be a consequence of the structure of $K(\mathbf{P})$ over $K(X)$. We begin by recalling some facts about direct images. Then we construct the canonical Koszul complex on $\mathbf{P}$. Koszul complexes, in various forms, give relations in the K-groups. For example, the first relation for $[\mathcal{O}_{\mathbf{P}}(1)]$ is precisely the relation satisfied by ℓ in $K(X)_e$ as in Chapter II.

Direct Images

For any coherent sheaf $\mathcal{F}$ on $\mathbf{P}$ we have the direct image $f_*\mathcal{F}$, which is coherent on X. We also have the higher direct image sheaves $R^if_*\mathcal{F}$, which are coherent $\mathcal{O}_X$-modules. They may be defined to be the sheaves associated to the presheaves

$$U \mapsto H^i(f^{-1}(U), \mathcal{F}).$$

In particular, $R^0f_*\mathcal{F} = f_*\mathcal{F}$. If $X = \mathrm{Spec}(A)$ is affine, then

$$R^if_*\mathcal{F} = H^i(\mathbf{P}, \mathcal{F})^{\sim}.$$

We record the following properties.

R 1. For any exact sequence

$$0 \to \mathcal{F}' \to \mathcal{F} \to \mathcal{F}'' \to 0$$

there is a (functorial) long exact sequence

$$0 \to f_*\mathcal{F}' \to f_*\mathcal{F} \to f_*\mathcal{F}'' \to R^1f_*\mathcal{F}' \to \cdots,$$

R 2. The cohomology functor has dimension $\leqq r$, that is

$$R^if_*(\mathcal{F}) = 0 \text{ for } i > r.$$

R 3. **(Projection Formula for R^if_*).** For $\mathcal{G} \in \mathfrak{B}_X$ we have

$$R^if_*(\mathcal{F} \otimes f^*\mathcal{G}) = R^if_*(\mathcal{F}) \otimes \mathcal{G}, \qquad \text{all} \quad i \geqq 0.$$

R 4. **(Serre's Theorem).** For all $\mathcal{F}$ there is n_0 such that

$$R^if_*(\mathcal{F}(n)) = 0 \qquad \text{for} \quad i > 0 \quad \text{and} \quad n \geqq n_0.$$

R 5. For $n \geqq 0$ and $\mathcal{M}$ coherent on X, we have

$$f_*(\mathcal{O}(n) \otimes f^*\mathcal{M}) = \operatorname{Sym}^n(\mathcal{E}) \otimes \mathcal{M}.$$

R 6. $R^i f_*(\mathcal{O}(n) \otimes f^*\mathcal{M}) = 0$ for $0 < i < r$, all $n \in \mathbf{Z}$;
for $i = r$, $n \geqq -r$;
and all $\mathcal{M}$ coherent on X.

The proofs can be found in [H], III, except for **R 2**, **R 5**, **R 6** which are easy consequences of [H], III. The properties are local on X, so we may assume $X = \operatorname{Spec}(A)$ where A is Noetherian.

For **R 2** we note that $\mathbf{P} = \mathbf{P}^r_A$ is covered by $r + 1$ affine open sets. By Leray's theorem [H], III, 4.5 the Cech cohomology with respect to this covering is the same as the ordinary cohomology for any coherent sheaf, and so vanishes in dimension $> r$.

As for **R 5** and **R 6**, they are special cases of the fact that the projection formula **R 3** is valid for any coherent $\mathcal{M}$ on X when $\mathcal{F} = \mathcal{O}(n)$, that is

R 7. $R^i f_*(\mathcal{O}(n) \otimes f^*\mathcal{M}) = R^i f_* \mathcal{O}(n) \otimes \mathcal{M}$ for $i \geqq 0, n \in \mathbf{Z}$.

The values of $R^i f_* \mathcal{O}(n)$ are given locally as the cohomology of $\mathcal{O}(n)$. The explicit computation of [H], III, 5.1 for the cohomology on projective space over an affine base without the extra $\mathcal{M}$ works just as well with an $\mathcal{M}$ to give the statements listed above.

Remark. **R 5** is valid for all $n \in \mathbf{Z}$ if it is understood that $\operatorname{Sym}^n(\mathcal{E}) = 0$ for $n < 0$. Then $R^r f_*(\mathcal{O}(n))$ is determined for all $n \in \mathbf{Z}$ by the existence of a duality

$$R^r f_* \mathcal{O}(n) \times R^0 f_* \mathcal{O}(-r - 1 - n) \to \Lambda^{r+1} \mathcal{E}$$

for all $n \in \mathbf{Z}$; and **R 6** follows from **R 5** and this duality. We do not need these further results, however.

The Koszul Complex on P

We are going to construct a **canonical resolution of** $\mathcal{O}_{\mathbf{P}}$. From the canonical surjection of $f^*\mathcal{E}$ onto $\mathcal{O}(1)$, we get a surjection

$$f^*\mathcal{E} \otimes \mathcal{O}(-1) \xrightarrow{d_1} \mathcal{O}_{\mathbf{P}} \to 0.$$

The sheaf $f^*\mathcal{O}(-1)$ is locally free, and we can therefore construct the Koszul complex as in Chapter IV, §2. Since for any invertible sheaf $\mathcal{G}$ and any locally free sheaf $\mathcal{F}$ we have an isomorphism

$$\Lambda^p(\mathcal{F} \otimes \mathcal{G}) \approx \Lambda^p \mathcal{F} \otimes \mathcal{G}^{\otimes p},$$

and since f^* commutes with Λ^p, we obtain the exact sequence which will be called **the Koszul complex on P**, or **Koszul resolution**

$$(2.1)\quad 0 \to f^*\Lambda^{r+1}\mathscr{E}(-r-1) \to f^*\Lambda^r\mathscr{E}(-r) \to \cdots \to f^*\mathscr{E}(-1) \to \mathscr{O}_{\mathbf{P}} \to 0.$$

For any coherent sheaf $\mathscr{F}$ on $\mathbf{P}$ we tensor the dual of the Koszul complex with $\mathscr{F}$ to obtain an exact sequence

$$(2.2)\quad 0 \to \mathscr{F} \to f^*\mathscr{E}^\vee \otimes \mathscr{F}(1) \to f^*\Lambda^2\mathscr{E}^\vee \otimes \mathscr{F}(2) \to \cdots \to f^*\Lambda^{r+1}\mathscr{E}^\vee \otimes \mathscr{F}(r+1) \to 0$$

which we denote by $\mathrm{Kos}^\vee(\mathscr{F})$. Then

$$\mathscr{F} \mapsto \mathrm{Kos}^\vee(\mathscr{F})$$

is exact on the category $\mathfrak{V}_{\mathbf{P}}$ of locally free sheaves on $\mathbf{P}$.

Regular Sheaves

We shall now analyze $K(\mathbf{P})$ by considering a subcategory of $\mathfrak{V}_{\mathbf{P}}$ which generates $K(\mathbf{P})$ and behaves particularly well under direct images.

Following Castelnuovo and Mumford, we say that a coherent sheaf $\mathscr{F}$ on $\mathbf{P}$ is **regular** if

$$R^i f_* \mathscr{F}(-i) = 0 \qquad \text{for} \quad i > 0.$$

Note that by **R 6**, if $\mathscr{M}$ is coherent on X, then $f^*\mathscr{M}$ is regular, and more generally, $\mathscr{O}(n) \otimes f^*\mathscr{M}$ is regular for $n \geqq 0$. If

$$0 \to \mathscr{F}' \to \mathscr{F} \to \mathscr{F}'' \to 0$$

is a short exact sequence of coherent sheaves, and $\mathscr{F}'$, $\mathscr{F}''$ are regular, it follows from the long exact sequence of **R 1** that $\mathscr{F}$ is regular. We let:

$\mathfrak{R}_{\mathbf{P}}$ = category of regular locally free sheaves on $\mathbf{P}$;

$K(\mathfrak{R}_{\mathbf{P}})$ = Grothendieck group of $\mathfrak{R}_{\mathbf{P}}$.

Proposition 2.1. *The inclusion of $\mathfrak{R}_{\mathbf{P}}$ in the category of all locally free sheaves $\mathfrak{V}_{\mathbf{P}}$ induces an isomorphism*

$$K(\mathfrak{R}_{\mathbf{P}}) \xrightarrow{\approx} K(\mathfrak{V}_{\mathbf{P}}) = K(\mathbf{P}).$$

Proof. We consider an auxiliary category. Let n be an integer $\geqq 0$. We let:

$\mathfrak{R}_n$ = category of elements $\mathscr{F} \in \mathfrak{V}_{\mathbf{P}}$ such that

$$R^i f_*(\mathscr{F}(j)) = 0 \qquad \text{for all} \quad i > 0 \quad \text{and} \quad j \geqq n - i.$$

In a short exact sequence of coherent sheaves as above, if $\mathscr{F}'$, $\mathscr{F}''$ are in $\mathfrak{R}_n$ then $\mathscr{F}$ is in $\mathfrak{R}_n$. We have $\mathfrak{R}_n \subset \mathfrak{R}_{n+1}$. In Proposition 2.2(1) we shall see that $\mathfrak{R}_n = \mathfrak{R}_{\mathbf{P}}$ for all n. Here we prove:

The inclusion of $\mathfrak{R}_n$ in $\mathfrak{R}_{n+1}$ induces an isomorphism

$$K(\mathfrak{R}_n) \xrightarrow{\approx} K(\mathfrak{R}_{n+1}).$$

Proof. Let $\mathscr{F} \in \mathfrak{R}_{n+1}$. Then by the definitions and **R 4**, it follows that $f^*\Lambda^p\mathscr{E}^\vee \otimes \mathscr{F}(p)$ is in $\mathfrak{R}_n$ for $p \geqq 1$. Then

$$[\mathscr{F}] \mapsto \sum (-1)^p [f^*\Lambda^p\mathscr{E}^\vee \otimes \mathscr{F}(p)]$$

is an additive map from $K(\mathfrak{R}_{n+1})$ to $K(\mathfrak{R}_n)$ because $\mathscr{F} \mapsto \mathrm{Kos}^\vee(\mathscr{F})$ is exact on $\mathfrak{V}_{\mathbf{P}}$, and this map gives the inverse of the natural homomorphism induced by the inclusion. By **R 4**, any locally free sheaf is in $\mathfrak{R}_n$ for sufficiently large n, whence we obtain the isomorphism

$$K(\mathfrak{R}_n) \xrightarrow{\approx} K(\mathbf{P})$$

for all n. Since $\mathfrak{R}_0$ is contained in $\mathfrak{R}_{\mathbf{P}}$, the proposition follows.

Remark. For future use, we define another category:

$\mathfrak{R}'_n$ = category of elements $\mathscr{F} \in \mathfrak{V}_{\mathbf{P}}$ such that

$$R^i f_*(\mathscr{F}(n + j)) = 0 \qquad \text{for all} \quad i > 0 \quad \text{and} \quad j \geqq 0.$$

Then we have the same statement as for $\mathfrak{R}_n$:

The inclusion $\mathfrak{R}'_n \to \mathfrak{V}_{\mathbf{P}}$ induces an isomorphism

$$\mathrm{K}(\mathfrak{R}'_n) \xrightarrow{\approx} K(\mathbf{P}).$$

The proof is the same as for $\mathfrak{R}_n$.

Proposition 2.2.

(1) *If* $\mathscr{F}$ *is regular, then* $\mathscr{F}(n)$ *is regular for all* $n \geqq 0$. *In particular,* $R^i f_*(\mathscr{F}) = 0$ *for* $i > 0$.

(2) *If* $\mathscr{F}$ *is regular, then the canonical homomorphism*

$$f^* f_* \mathscr{F} \to \mathscr{F}$$

is surjective, and if $\mathscr{Z}$ *is its kernel, then* $\mathscr{Z}(1)$ *is regular.*

(3) *If* $\mathscr{F}$ *is regular and in* $\mathfrak{V}_{\mathbf{P}}$ *then* $f_* \mathscr{F}$ *is in* $\mathfrak{V}_X$.

Quillen [Q], §8 has given an elegant proof of this proposition. We include his proof, since we shall need some of the concepts later.

Decompose the Koszul resolution into short exact sequences

$$0 \to \mathscr{K}_p \to f^* \Lambda^p \mathscr{E}(-p) \to \mathscr{K}_{p-1} \to 0 \tag{2.3}$$

of locally free sheaves, with

$$\mathscr{K}_0 = \mathscr{O}_{\mathbf{P}} \qquad \text{and} \qquad \mathscr{K}_r = f^* \Lambda^{r+1} \mathscr{E}(-r-1).$$

We prove (1). Let $\mathscr{F}$ be regular in $\mathfrak{V}_{\mathbf{P}}$. Tensoring the short exact sequence with $\mathscr{F}(p)$ gives the short exact sequence

$$0 \to \mathscr{K}_p \otimes \mathscr{F}(p) \to f^* \Lambda^p \mathscr{E} \otimes \mathscr{F} \to \mathscr{K}_{p-1} \otimes \mathscr{F}(p) \to 0. \tag{2.4}$$

It suffices to prove that $\mathscr{F}(1)$ is regular. We shall now prove by descending induction on p that $\mathscr{K}_p \otimes \mathscr{F}(p+1)$ is regular for all p. Let first $p = r + 1$ so $\mathscr{K}_{r+1} = 0$. For $p = r$, by the projection formula **R 3**, we have

$$\begin{aligned} R^i f_*(\mathscr{K}_r \otimes \mathscr{F}(r+1)(-i)) &= R^i f_*(f^* \Lambda^{r+1} \mathscr{E} \otimes \mathscr{F}(-i)) \\ &= \Lambda^{r+1}(\mathscr{E}) \otimes R^i f_* \mathscr{F}(-i) \\ &= 0 \qquad \text{for} \quad i > 0. \end{aligned}$$

This proves that $\mathscr{K}_r \otimes \mathscr{F}(r+1)$ is regular. For the inductive step, we use the long exact sequence

$$R^i f_*(f^* \Lambda^p \mathscr{E} \otimes \mathscr{F}(-i)) \to R^i f_*(\mathscr{K}_{p-1} \otimes \mathscr{F}(p-i)) \to R^{i+1} f_*(\mathscr{K}_p \otimes \mathscr{F}(p-i)).$$

The term on the far right is 0 by induction. The term on the left is 0 by projection formula **R 3** and the hypothesis that $\mathscr{F}$ is regular. This proves that $\mathscr{K}_{p-1} \otimes \mathscr{F}(p)$ is regular. For $p = 0$ this shows that

$$\mathscr{F}(1) = \mathscr{K}_0 \otimes \mathscr{F}(1)$$

is regular, thus proving (1).

We prove (2). We tensor the first short exact sequence of the Koszul resolution of $\mathcal{O}_{\mathbf{P}}$ with $\mathscr{F}(n)$ to get the exact sequence

$$0 \to \mathscr{K}_1 \otimes \mathscr{F}(n) \to f^*\mathscr{E} \otimes \mathscr{F}(n-1) \to \mathscr{F}(n) \to 0.$$

In the proof of (1), we have just seen that $\mathscr{K}_1 \otimes \mathscr{F}(2)$ is regular. By (1) and the definition of regularity, it follows that $R^1f_*(\mathscr{K}_1 \otimes \mathscr{F}(n)) = 0$ for $n \geqq 1$, and therefore

$$\mathscr{E} \otimes f_*(\mathscr{F}(n-1)) \to f_*\mathscr{F}(n) \to 0$$

is surjective for $n \geqq 1$. We have a commutative diagram

$$\begin{array}{ccc} \mathscr{E} \otimes \mathrm{Sym}^{n-1}(\mathscr{E}) \otimes f_*\mathscr{F} & \longrightarrow & \mathrm{Sym}^n(\mathscr{E}) \otimes f_*\mathscr{F} \\ \downarrow & & \downarrow \\ \mathscr{E} \otimes f_*\mathscr{F}(n) & \longrightarrow & f_*\mathscr{F}(n+1). \end{array}$$

By induction, the left vertical arrow can be assumed to be surjective, and the bottom arrow is surjective so the right arrow is surjective. Taking direct sums, we get a surjection

$$\mathrm{Sym}(\mathscr{E}) \otimes f_*\mathscr{F} \to \bigoplus_{n=0}^{\infty} f_*\mathscr{F}(n) \to 0.$$

The assertion of (2) is local on the base, so without loss of generality we may assume that X is affine, $X = \mathrm{Spec}(A)$, and $\mathbf{P} = \mathbf{P}^r_A$. Then the direct sum on the right is usually denoted by

$$\bigoplus_{n=0}^{\infty} f_*\mathscr{F}(n) = (\Gamma_*\mathscr{F})^\sim = \bigoplus_{n=0}^{\infty} H^0(\mathbf{P}, \mathscr{F}(n))^\sim.$$

Cf. [H], II, §5, especially 5.15. Now we use a general fact which we state as follows.

Let R be a graded ring. We suppose that R_0 is Noetherian, R_1 is a finitely generated R_0-module, and R is finitely generated as R_0-algebra by R_1. We let

$$X = \mathrm{Spec}(R_0) \qquad \text{and} \qquad \mathbf{P} = \mathrm{Proj}(R).$$

We let $f: \mathbf{P} \to X$ be the structural morphism. Let M be a graded R-module. Then we have $M^{\sim}$ (projective tilde), which is a quasi-coherent sheaf on $\mathbf{P}$. We define two graded modules M, M' to be **quasi equal** if $M_n = M'_n$ for all n sufficiently large. We say that M is **quasi-finitely generated over** R if M is quasi equal to a finitely generated graded module over R. The result we have in mind is then as follows.

The association $M \mapsto M^{\sim}$ is an equivalence of categories between the quasi-finitely generated graded modules over R (modulo quasi-equality) and the category of coherent $\mathcal{O}_{\mathbf{P}}$-modules. Furthermore the functors

$$M \mapsto M^{\sim}$$

and

$$\mathscr{F} \mapsto \Gamma_* \mathscr{F} = \bigoplus_{n=0}^{m} H^0(\mathbf{P}, \mathscr{F}(n))$$

are inverse to each other (up to isomorphism). Finally, if N is a finitely generated R_0-module, and $N^{\sim}$ its corresponding coherent sheaf on $X =$ $\operatorname{Spec}(R_0)$ (affine tilde in this case), then there is a natural isomorphism

$$(R \otimes_{R_0} N)^{\sim} \xrightarrow{\approx} f^*(N^{\sim}) = \mathcal{O}_{\mathbf{P}} \otimes_{\mathcal{O}_X} N^{\sim}.$$

Note that the tilde on the left is the projective tilde, and the tilde on the right is the affine tilde.

It is really not the place here to reproduce a complete proof of the above elementary theorem. Cf. [H], II, Proposition 5.15 and Exercise 5.9 (where a reference to a field k is unnecessary), stemming from Serre's original *Faisceaux Algébriques Cohérents.*

We then apply the theorem to the case when $R = \operatorname{Sym}(E)$, $\mathscr{E} = E^{\sim}$, to conclude the proof of the first assertion in (2), that

$$f^* f_* \mathscr{F} \to \mathscr{F} \to 0$$

is exact.

As to the second assertion, let $\mathscr{L}$ be the kernel. We look at the beginning of the long cohomology sequence:

$$R^0 f_*(f^* f_* \mathscr{F}) \to R^0 f_*(\mathscr{F}) \to R^1 f_*(\mathscr{L}) \to R^1 f_*(f^* f_* \mathscr{F}).$$

The term on the far right is 0 by **R 6**. The term on the far left is just $f_* \mathscr{F}$ by **R 5**, and the arrow on the left is the identity. Hence $R^1 f_*(\mathscr{L}) = 0$. For $i \geqq 2$, we consider the short exact sequence

$$0 \to \mathscr{L}(1-i) \to f^* f_* \mathscr{F}(1-i) \to \mathscr{F}(1-i) \to 0$$

giving rise to the long exact sequence

$$R^{i-1}f_*(\mathscr{F}(1-i)) \to R^i f_*(\mathscr{Z}(1-i)) \to R^i f_*(f^*f_*\mathscr{F}(1-i)).$$

The term on the left is 0 because $\mathscr{F}$ is regular. The term on the right is 0 by **R 6**. This concludes the proof that $\mathscr{Z}(1)$ is regular, and thus concludes the proof of (2).

Finally we prove (3). We let $\mathscr{F}$ be regular locally free. By (1) each term to the right of $\mathscr{F}$ in the Koszul complex $\mathrm{Kos}^\vee(\mathscr{F})$ has vanishing higher direct images. Therefore applying f_* to $\mathrm{Kos}^\vee(\mathscr{F})$ yields an exact sequence

$$0 \to f_*\mathscr{F} \to \Lambda^1\mathscr{E}^\vee \otimes f_*\mathscr{F}(1) \to \cdots \to \Lambda^{r+1}\mathscr{E}^\vee \otimes f_*\mathscr{F}(r+1) \to 0.$$

By the remarks in the introduction to this chapter, to show $f_*\mathscr{F}$ locally free, it suffices to show that $f_*\mathscr{F}(i)$ is locally free for $i = 1,\ldots,r+1$. By (1), $\mathscr{F}(i)$ is regular for $i \geqq 1$. Moving to the right it suffices to show that $f_*\mathscr{F}(n)$ is locally free for sufficiently large n. But this is a general fact:

Let $\mathscr{F}$ be a locally free sheaf on $\mathbf{P}$. *Then $f_*\mathscr{F}(n)$ is locally free for sufficiently large n.*

Proof. The statement is local in X, so we may assume that X is affine, and the statement is equivalent to the fact that $f_*\mathscr{F}(n)$ is a flat $\mathscr{O}_X$-module for large n. The scheme $\mathbf{P}$ is just projective space over the affine X. Then $\mathscr{F}(m)$ is generated by global sections for m large, so for m large there is a surjection from a finite sum of $\mathscr{O}_{\mathbf{P}}$ onto $\mathscr{F}(m)$. Twisting back by $-m$, we obtain an exact sequence

$$0 \to \mathscr{F}' \to \mathscr{G} \to \mathscr{F} \to 0,$$

where $\mathscr{G}$ is a direct sum of sheaves of type $\mathscr{O}(-m)$. Twisting by $n > m$ and tensoring with $f^*\mathscr{M}$ where $\mathscr{M}$ is coherent of X, we get a short exact sequence

$$0 \to \mathscr{F}'(n) \otimes f^*\mathscr{M} \to \mathscr{G}(n) \otimes f^*\mathscr{M} \to \mathscr{F}(n) \otimes f^*\mathscr{M} \to 0,$$

whence the exact cohomology sequence

$$R^i f_*(\mathscr{G}(n) \otimes f^*\mathscr{M}) \to R^i f_*(\mathscr{F}(n) \otimes f^*\mathscr{M}) \to R^{i+1} f_*(\mathscr{F}'(n) \otimes f^*\mathscr{M}).$$

Let $i > r$. Then the term on the right is 0 because the cohomology functor has dimension $\leqq r$ by **R 2**. The term on the left is 0 for $n > n_0$

by **R 6**, so the term in the middle is 0. Then we can do a descending induction to prove that given $\mathscr{F}$ there exists n_1 such that

$$R^i f_*(\mathscr{F}(n) \otimes f^*\mathscr{M}) = 0 \qquad \text{for} \quad i > 0, \quad n > n_1$$

and all coherent $\mathscr{M}$ on X.

Taking $n > n_1$ we obtain a commutative diagram with exact rows:

$$\begin{array}{ccccccc}
 & f_*(\mathscr{F}'(n)) \otimes \mathscr{M} & \longrightarrow & f_*(\mathscr{G}(n)) \otimes \mathscr{M} & \longrightarrow & f_*(\mathscr{F}(n)) \otimes \mathscr{M} & \longrightarrow 0 \\
 & \downarrow u & & \downarrow v & & \downarrow w & \\
0 \longrightarrow & f_*(\mathscr{F}'(n) \otimes f^*\mathscr{M}) & \longrightarrow & f_*(\mathscr{G}(n) \otimes f^*\mathscr{M}) & \longrightarrow & f_*(\mathscr{F}(n) \otimes f^*\mathscr{M}) & \longrightarrow 0
\end{array}$$

But **R 5** implies that v is an isomorphism. It follows that w is surjective for $n > n_1$. Then u is surjective for $n > n_2$ by applying this result to $\mathscr{F}'$ instead of $\mathscr{F}$. The snake lemma implies that w is injective, so w is an isomorphism. Therefore the functor

$$\mathscr{M} \mapsto f_*(\mathscr{F}(n)) \otimes \mathscr{M}$$

is an exact functor, so $f_*\mathscr{F}(n)$ is flat. This concludes the proof of (3), and also the proof of Proposition 2.2.

Remark. This final general fact is an elementary result of algebraic geometry which is proved in the course of proving [H], III, 9.9, (i) implies (ii). No hypothesis about A being without divisors of zero is used in that part of the proof, which uses the Cech complex directly. We included another proof for the convenience of the reader, because it fitted the techniques used in this section.

The next result can be viewed as a sheaf version of the fact that $K(\mathfrak{N}_{\mathbf{P}})$ is generated as a module over $K(X)$ by $1, \ell^{-1}, \ldots, \ell^{-r}$. We use the preceding proposition to show:

Any regular sheaf $\mathscr{F}$ on **P** *has a* **canonical resolution**

$$\text{(2.5)} \qquad \begin{aligned} 0 \to f^*\mathscr{T}_r(\mathscr{F})(-r) \to f^*\mathscr{T}_{r-1}(\mathscr{F})(-r+1) \to \cdots \\ \to f^*\mathscr{T}_1(\mathscr{F})(-1) \to f^*\mathscr{T}_0(\mathscr{F}) \to \mathscr{F} \to 0, \end{aligned}$$

where $\mathscr{T}_i(\mathscr{F})$ are sheaves on X, and the functors $\mathscr{F} \mapsto \mathscr{T}_i(\mathscr{F})$ are exact.

This is constructed inductively as follows. Let $\mathscr{T}_0(\mathscr{F}) = f_*\mathscr{F}$, and let $\mathscr{Z}_0$ be the kernel of the canonical map from $f^*\mathscr{T}_0\mathscr{F}$ to $\mathscr{F}$. Then $\mathscr{Z}_0(1)$ is regular by (2), so we may define $\mathscr{T}_1(\mathscr{F})$ and $\mathscr{Z}_1$ by

$$\mathscr{T}_1(\mathscr{F}) = f_*(\mathscr{Z}_0(1)),$$

and the exact sequence

$$0 \to \mathscr{Z}_1(1) \to f^*\mathscr{T}_1(\mathscr{F}) \to \mathscr{Z}_0(1) \to 0.$$

This gives an exact sequence

$$0 \to \mathscr{Z}_1 \to f^*\mathscr{T}_1(\mathscr{F})(-1) \to f^*\mathscr{T}_0(\mathscr{F}) \to \mathscr{F} \to 0$$

with $\mathscr{Z}_1(2)$ regular. Inductively define $\mathscr{T}_p(\mathscr{F})$, $\mathscr{Z}_p$ by

$$\mathscr{T}_p(\mathscr{F}) = f_*(\mathscr{Z}_{p-1}(p)), \tag{2.6}$$

$$0 \to \mathscr{Z}_p(p) \to f^*\mathscr{T}_p(\mathscr{F}) \to \mathscr{Z}_{p-1}(p) \to 0. \tag{2.7}$$

One sees by induction that $\mathscr{Z}_p(p+1)$ is regular, and that $\mathscr{T}_p$ and $\mathscr{Z}_p$ are exact functors of regular sheaves $\mathscr{F}$. In addition,

$$R^if_*(\mathscr{Z}_{i+p}(p)) = 0$$

for all $i \geqq 0$, $p \geqq 0$. For $i = 0$ this follows by applying f_* to the sequence defining $\mathscr{Z}_p(p)$. For $i > 0$ it follows by induction on i and the exact sequence

$$R^{i-1}f_*(\mathscr{Z}_{i+p-1}(p)) \to R^if_*(\mathscr{Z}_{i+p}(p)) \to R^if_*(f^*\mathscr{T}_{i+p}(\mathscr{F})(-i)).$$

In particular, $\mathscr{Z}_r(r)$ is regular because $R^if_* = 0$ for $i > r$. Since $f_*\mathscr{Z}_r(r) = 0$, we get $\mathscr{Z}_r(r) = 0$ by (2) of the proposition, so $\mathscr{Z}_r = 0$ as desired.

Remark. As shown easily in Quillen [Q], §8, the sheaves $\mathscr{T}_p(\mathscr{F})$ are uniquely determined by the following property. Let $\mathscr{G}_p$ be coherent sheaves on X such that there is a resolution

$$0 \to f^*\mathscr{G}_r(-r) \to f^*\mathscr{G}_{r-1}(-r+1) \to \cdots \to f^*\mathscr{G}_1(-1) \to f^*\mathscr{G}_0 \to \mathscr{F} \to 0.$$

Then $\mathscr{G}_p \approx \mathscr{T}_p(\mathscr{F})$ for $p = 0,\ldots,r$. We won't need this property, which can be viewed as a sheaf-theoretic version of the fact proved in Theorem 2.3 that $1, \ell^{-1},\ldots,\ell^{-r}$ form a basis of $K(\mathbf{P}(\mathscr{E}))$ over $K(X)$.

We state two other properties of the canonical resolution.

If $\mathscr{F}$ *is locally free as well as regular, then each* $\mathscr{T}_p(\mathscr{F})$ *is locally free on* X *and* $\mathscr{Z}_p$ *is locally free on* $\mathbf{P}$.

This follows from (3) of Proposition 2.2 and induction on p.

If we denote the canonical resolution by $\mathscr{C}(\mathscr{F})$, *and*

$$0 \to \mathscr{F}' \to \mathscr{F} \to \mathscr{F}'' \to 0$$

is an exact sequence of regular sheaves, then the sequence of canonical resolutions

$$0 \to \mathscr{C}(\mathscr{F}') \to \mathscr{C}(\mathscr{F}) \to \mathscr{C}(\mathscr{F}'') \to 0$$

is also exact.

This follows from the construction and is left to the reader.

Theorem 2.3. *Let* $e = [\mathscr{E}]$ *in* $K(X)$. *Then we have an isomorphism of* $K(X)$*-algebras*

$$K(X)_e \xrightarrow{\approx} K(\mathbf{P}(\mathscr{E}))$$

which sends the canonical generator ℓ *on* $[\mathscr{O}(1)]$.

Here $K(X)_e$ is the λ-ring extension of $K(X)$ described in Chapter I §2, with generator ℓ and relation

$$\sum_{i=0}^{r+1} (-1)^i \lambda^i(e) \ell^{r+1-i} = 0.$$

Proof. Let $\ell_0 = [\mathscr{O}(1)]$. The Koszul resolution shows that ℓ_0 satisfies the same relation as ℓ. Hence there is a unique homomorphism

$$\varphi \colon K(X)_e \to K(\mathbf{P})$$

mapping ℓ on ℓ_0. If $\mathscr{F}$ is a regular locally free sheaf, then the sheaves $\mathscr{T}_p(\mathscr{F})$ are locally free, and the canonical resolution (2.5) shows that

$$[\mathscr{F}] = \sum_{p=0}^{r} (-1)^p [f^* \mathscr{T}_p \mathscr{F}] \cdot \ell_0^{-p}.$$

By Proposition 2.1 such $[\mathscr{F}]$ generate $K(\mathbf{P})$, so φ is surjective.

Let $\mathfrak{R}'$ be the category of locally free sheaves $\mathscr{F}$ on $\mathbf{P}$ such that

$$R^i f_*(\mathscr{F}(j)) \qquad \text{for} \quad i > 0, \quad j \geqq 0.$$

By the remark after Proposition 2.1, $K(\mathfrak{R}') = K(\mathbf{P})$. The map

$$\psi: K(\mathfrak{R}') \to \bigoplus_{n=0}^{r} K(X)$$

given by

$$\psi([\mathscr{F}]) = ([f_*\mathscr{F}], [f_*\mathscr{F}(1)], \ldots, [f_*\mathscr{F}(r)])$$

is well defined because the functor

$$\mathscr{F} \mapsto f_*\mathscr{F}(n)$$

is exact on $\mathfrak{R}'$ for $n \geqq 0$, and then ψ is a homomorphism. Consider the composite

$$\bigoplus_{n=0}^{r} K(X) \overset{\approx}{\to} K(X)_e \overset{\varphi}{\to} K(\mathbf{P}) = K(\mathfrak{R}') \overset{\psi}{\to} \bigoplus_{n=0}^{r} K(X),$$

where the first isomorphism takes $\bigoplus_{n=0}^{r} a_n$ to $\sum_{n=0}^{r} a_n \ell^{-n}$. The composite is given by a triangular matrix with 1's on the diagonal, since for a locally free sheaf $\mathscr{A}_i$ on X,

$$f_*(f^*\mathscr{A}_i \otimes \mathscr{O}(-i) \otimes \mathscr{O}(j)) = \begin{cases} \mathscr{A}_i & \text{if } j = i, \\ 0 & \text{if } j < i. \end{cases}$$

Hence φ is injective as well as surjective. This proves Theorem 2.3.

We identify $K(\mathbf{P}(E))$ with $K(\mathfrak{R}_{\mathbf{P}})$ by Proposition 2.1. There exists a unique homomorphism

$$f_K: K(\mathbf{P}(\mathscr{E})) \to K(X)$$

such that

$$f_K[\mathscr{F}] = [f_*\mathscr{F}]$$

for any regular locally free sheaf $\mathscr{F}$ on $\mathbf{P}(\mathscr{E})$. Indeed note by Proposition 2.2 that $f_*\mathscr{F}$ is a locally free sheaf on X, and that $R^i f_*\mathscr{F} = 0$ for $i > 0$, so f_* is exact on $\mathfrak{R}$.

Corollary 2.4. *Under the isomorphism $K(\mathbf{P}(\mathscr{E})) \approx K(X)_e$, f_K corresponds to the functional f_e of Chapter* I, §2.

Proof. By construction of f_K, and **R 5**, **R 6**, we have

$$f_K[\mathscr{O}(n)] = [\operatorname{Sym}^n \mathscr{E}] \qquad \text{for} \quad n \geqq 0.$$

To complete the proof we must verify that $[\operatorname{Sym}^n \mathscr{E}] = \sigma^n e$, i.e. that

$$\left(\sum_{n=0}^{\infty} [\operatorname{Sym}^n \mathscr{E}] t^n\right)\left(\sum_{n=0}^{r+1} (-1)^n [\Lambda^n \mathscr{E}] t^n\right) = 1.$$

We give two proofs for this. On the one hand, there is a complex

$$0 \to \Lambda^{r+1}\mathscr{E} \otimes \operatorname{Sym}(\mathscr{E}) \to \Lambda^r\mathscr{E} \otimes \operatorname{Sym}(\mathscr{E}) \to \cdots \to \Lambda^0\mathscr{E} \otimes \operatorname{Sym}(\mathscr{E}) \to \mathscr{O}_X \to 0$$

which is the Koszul complex over the symmetric algebra $\operatorname{Sym}(\mathscr{E})$, with respect to the map which sends $\Lambda^1\mathscr{E}$ naturally in $\operatorname{Sym}(\mathscr{E})$:

$$d_1 : \Lambda^1\mathscr{E} \otimes \operatorname{Sym}(\mathscr{E}) \to \Lambda^0\mathscr{E} \otimes \operatorname{Sym}(\mathscr{E}) = \operatorname{Sym}(\mathscr{E}).$$

Locally, if $T_0, \ldots, T_r$ is a basis of the free module E over the ring R, then $\operatorname{Sym}(E) = R[T_0, \ldots, T_r]$, and d_1 maps a basis for a free module of rank $r+1$ over $\operatorname{Sym}(E)$ to the elements $T_0, \ldots, T_r$. Thus locally, the above complex is the Koszul complex of a regular sequence (namely the sequence of variables $T_0, \ldots, T_r$ in the ring $R[T_0, \ldots, T_r]$). Hence the Koszul complex is exact. Since d_p maps $\Lambda^p\mathscr{E} \otimes \operatorname{Sym}^q(\mathscr{E})$ into

$$\Lambda^{p-1}\mathscr{E} \otimes \operatorname{Sym}^{q+1}(E),$$

we can decompose this complex into a direct sum corresponding to graded component, and hence we have an exact sequence

$$0 \to \Lambda^{r+1}\mathscr{E} \otimes \operatorname{Sym}^{n-r-1}(\mathscr{E}) \to \cdots \to \Lambda^1\mathscr{E} \otimes \operatorname{Sym}^{n-1}(\mathscr{E}) \to \operatorname{Sym}^n(\mathscr{E}) \to 0$$

for integers $n \geqq 1$, it being understood that $\operatorname{Sym}^j(\mathscr{E}) = 0$ if $j < 0$. This last exact sequence gives precisely the desired relation in the K-group. Alternatively, we would use the splitting property which will be proved below, and which reduces the assertion to the case when $\mathscr{E}$ has rank 1, when it is obvious.

Corollary 2.5. $F^n K(\mathbf{P}(\mathscr{E})) \cap K(X) = F^n K(X)$.

Proof. Apply Theorem 1.2 of Chapter III.

Remark. The set of positive elements in $K(\mathbf{P}(\mathscr{E}))$, i.e. the classes of locally free sheaves, may be larger than the set $\mathbf{E}_e$ described formally in Chapter I, §2.

Lemma 2.6 (Projection Formula). *If* $x \in K(\mathbf{P}\mathscr{E})$, $y \in K(X)$, *then*

$$f_K(x \cdot f^K y) = f_K(x) \cdot y.$$

Proof. By Proposition 2.1 and linearity we may take $x = [\mathscr{F}]$ with $\mathscr{F}$ regular and locally free on $\mathbf{P}\mathscr{E}$, and $y = [\mathscr{G}]$, $\mathscr{G}$ locally free on X. Then $\mathscr{F} \otimes f^*\mathscr{G}$ is regular by **R 6**, and

$$f_*(\mathscr{F} \otimes f^*\mathscr{G}) \approx f_*(\mathscr{F}) \otimes \mathscr{G}$$

which is the required formula.

Theorem 2.7 (Splitting Property). *Given a locally free sheaf* $\mathscr{E}$ *on* X, *there is a morphism* $f: X' \to X$ *such that*

$$f^K: K(X) \to K(X')$$

is injective,

$$F^n K(X') \cap K(X) = F^n K(X)$$

for all n, *and*

$$f^K[\mathscr{E}] = [\mathscr{L}_1] + \cdots + [\mathscr{L}_m]$$

for some invertible sheaves $\mathscr{L}_i$ *on* X'.

Proof. First let f be a bundle projection $\mathbf{P}(\mathscr{E}) \to X$. Then we have the tautological exact sequence

$$0 \to \mathscr{H} \to f^*\mathscr{E} \to \mathscr{O}(1) \to 0$$

so $[f^*\mathscr{E}] = [\mathscr{H}] + [\mathscr{O}(1)]$, and the rank of $\mathscr{H}$ is one less than the rank of $\mathscr{E}$. By induction, we take a sequence of such bundle projections to conclude the proof.

V §3. Grothendieck and Topological Filtrations

In this section we assume X *is a connected, Noetherian scheme with an ample invertible sheaf* $\mathscr{L}$. Recall that this means that for any coherent sheaf $\mathscr{F}$ on X there is an integer $n_0 = n_0(\mathscr{F})$ such that for all $n \geqq n_0$, $\mathscr{F} \otimes \mathscr{L}^{\otimes n}$ is generated by its global sections. For example, if X is quasi-projective over an affine Noetherian base scheme S, then X has an ample invertible sheaf; for any ample $\mathscr{L}$ on X, and sufficiently large n, there is a locally closed imbedding i of X in some $\mathbf{P}_S^N$ such that $\mathscr{L}^{\otimes n} = i^*\mathcal{O}(1)$ (cf. [H], II, 7.6). We now connect ample sheaves with ample elements as defined in Chapter III before Lemma 1.4.

Lemma 3.1. *If $\mathscr{L}$ is ample on X, then $u = [\mathscr{L}]$ is an ample line element for $K(X)$.*

Proof. Given $x \in K(X)$, we must show that

$$u^n x = e - m$$

for some positive integers n, m, and a positive e in $K(X)$ (Chapter III, §1). Choose locally free sheaves $\mathscr{E}_1$, $\mathscr{E}_2$ with

$$x = [\mathscr{E}_1] - [\mathscr{E}_2].$$

For large n there is an $m > 0$ and a surjection

$$\mathcal{O}_X^{\oplus m} \xrightarrow{\alpha} \mathscr{E}_2 \otimes \mathscr{L}^{\otimes n} \longrightarrow 0.$$

If $\mathscr{E}'$ is the kernel of α, then

$$u^n x = [\mathscr{E}_1 \otimes \mathscr{L}^{\otimes n}] + [\mathscr{E}'] - [\mathcal{O}_X^{\oplus m}] = e - m$$

as required.

As in Chapter III, §1 let $F^n K(X)$ denote the γ-filtration on the λ-ring $K(X)$. We modify slightly a definition of H. Bass (also used in SGA 6 and Manin), and introduce another filtration, denoted $F_{\text{top}}^n K(X)$. For this some notation will be useful.

In this section a **complex** $\mathscr{E}^{\cdot}$ on X will be a bounded complex of locally free sheaves

$$0 \longrightarrow \mathscr{E}^a \xrightarrow{d^a} \mathscr{E}^{a+1} \longrightarrow \cdots \longrightarrow \mathscr{E}^b \longrightarrow 0.$$

We say that $\mathscr{E}^{\cdot}$ **represents** an element $x \in K(X)$ if

$$x = \sum_{i=a}^{b} (-1)^i [\mathscr{E}^i].$$

The **support** of $\mathscr{E}^{\cdot}$, denoted $|\mathscr{E}^{\cdot}|$, will be the set of points $x \in X$ at which the induced complex of vector spaces $\mathscr{E}^{\cdot}(x)$:

$$0 \to \mathscr{E}^a(x) \to \cdots \to \mathscr{E}^b(x) \to 0$$

over the residue field $\kappa(x) = \mathcal{O}_{x,X}/\mathcal{M}_{x,X}$ is not exact; $|\mathcal{E}^{\cdot}|$ is the union of the supports of the homology sheaves $\mathcal{H}^i(\mathcal{E}^{\cdot})$, so is a closed subset of X.

If Z is an irreducible closed subset of a Noetherian scheme Y, recall that the **codimension** of Z in Y, denoted codim(Z, Y) is the greatest length of a chain of irreducible closed subsets.

$$Z = V_0 \subsetneqq V_1 \subsetneqq \cdots \subsetneqq V_d \subset Y.$$

For an arbitrary closed $Z \subset Y$, codim(Z, Y) is defined to be the smallest of the codimensions of the irreducible components of Z in Y. With this definition, if $Z \subset Y \subset X$, then

$$\text{(3.1)} \qquad \operatorname{codim}(Z, Y) + \operatorname{codim}(Y, X) \leqq \operatorname{codim}(Z, X).$$

If $Z = \emptyset$, codim$(Z, Y) = +\infty$. The **dimension** of Y, dim(Y), is the maximum codimension of any non-empty closed subset. We define:

$F^n_{\text{top}}K(X) =$ set of elements $x \in K(X)$ such that for any finite family of closed subsets $\{Y_\alpha\}$ of X, x can be represented by a complex $\mathcal{E}^{\cdot}$ on X such that for any finite family of closed subsets $\{Y_\alpha\}$ of X, x can be represented by a complex $\mathcal{E}^{\cdot}$ on X such that

$$\operatorname{codim}(|\mathcal{E}^{\cdot}| \cap Y_\alpha, Y_\alpha) \geqq n$$

for all α. We say that such $\mathcal{E}^{\cdot}$ **represents** x **with respect to** $\{Y_\alpha\}$ **and** n.

We may write $F^n_{\text{top}}X$, or simply F^n_{top}, for $F^n_{\text{top}}K(X)$, and call it the **topological filtration**.

Proposition 3.2.

(a) *The $F^n_{\text{top}}K(X)$ define a ring filtration on $K(X)$.*

(b) *If* $\dim(X) \leqq d$, *then* $F^{d+1}_{\text{top}}K(X) = 0$.

Proof. We show first that F^n_{top} is an additive subgroup of $K(X)$. Given $x, y \in F^n_{\text{top}}$, to show $x - y \in F^n_{\text{top}}$, suppose closed subsets Y_α are specified. Choose complexes $\mathcal{E}^{\cdot}$ (resp. $\mathcal{F}^{\cdot}$) representing x (resp. y) with respect to $\{Y_\alpha\}$ and n. Then

$$\mathcal{E}^{\cdot} \oplus \mathcal{F}^{\cdot}[-1]$$

represents $x - y$ with respect to $\{Y_\alpha\}$ and n, where $\mathcal{F}^{\cdot}[-1]$ denotes the *shift* of $\mathcal{F}^{\cdot}$: $\mathcal{F}^{\cdot}[-1]^k = \mathcal{F}^{k-1}$.

To finish the proof of (a), we must show that if $x \in F^m_{\text{top}}$, $y \in F^n_{\text{top}}$, then $x \cdot y \in F^{m+n}_{\text{top}}$. Given closed subsets Y_α, chose $\mathcal{E}^{\cdot}$ representing x with respect to $\{Y_\alpha\}$ and m. Then choose $\mathcal{F}^{\cdot}$ representing y with respect to

$\{|\mathscr{E}^{\cdot}| \cap Y_\alpha\}$ and n. It follows that $\mathscr{E}^{\cdot} \otimes \mathscr{F}^{\cdot}$ represents $x \cdot y$ with respect to $\{Y_\alpha\}$ and $m + n$. The point here is that $\mathscr{E}^{\cdot} \otimes \mathscr{F}^{\cdot}$ is exact where *either* $\mathscr{E}^{\cdot}$ or $\mathscr{F}^{\cdot}$ is exact, so

$$|\mathscr{E}^{\cdot} \otimes \mathscr{F}^{\cdot}| = |\mathscr{E}^{\cdot}| \cap |\mathscr{F}^{\cdot}|;$$

the condition on codimension follows from (3.1).

For (b), take $Y = X$; if $x \in F_{\text{top}}^{d+1}X$, x is represented by a complex $\mathscr{E}^{\cdot}$ exact on all of X, so

$$x = \sum (-1)^i [\mathscr{E}^i] = 0$$

by definition of $K(X)$.

For $\mathscr{L}$ ample on X, and $\mathscr{F}$ any coherent sheaf on X, write as usual $\mathscr{F}(n)$ for $\mathscr{F} \otimes \mathscr{L}^{\otimes n}$, and write $\mathscr{F}(x)$ for the fiber of F over the residue field $\kappa(x)$.

Lemma 3.3. *Given a surjection $\mathscr{F} \to \mathscr{G}$ of coherent sheaves on X, and a finite set S of points in X such that $\mathscr{G}(x) \neq 0$ for all $x \in S$, then for all sufficiently large n there is a section of $\mathscr{F}(n)$ whose image in $\mathscr{G}(n)(x)$ is not zero for any $x \in S$.*

Proof. For large n_0 there is a section f of $\mathscr{L}^{\otimes n_0}$ such that the complement X_f of its zero-scheme is affine, and $S \subset X_f$ (cf. [H], pp. 154–155, or [EGA], II, 4.5.4). On X_f, $\mathscr{F} \to \mathscr{G}$ corresponds to a surjection of modules, from which one sees that there is a t in $\Gamma(X_f, \mathscr{F})$ whose image in $\mathscr{G}(x)$ is non-zero for $x \in S$. For large m, $f^m t$ extends to a section of $\mathscr{F}(mn_0)$. For all large p there are sections g of $\mathscr{L}^{\otimes p}$ that are not zero at $x \in S$. Then $gf^m t$ is a section of $\mathscr{F}(n)$ as required, with $n = mn_0 + p$.

Lemma 3.4. *Let $\mathscr{E}_1, \ldots, \mathscr{E}_p$ be locally free sheaves of the same rank r on X, and S a finite set of points of X. Then there are integers $m_1, \ldots, m_r$, and homomorphisms*

$$\mathscr{L}^{\otimes m_1} \oplus \cdots \oplus \mathscr{L}^{\otimes m_r} \to \mathscr{E}_i \tag{3.2}$$

for $i = 1, \ldots, p$, whose fibres at each $x \in S$ are isomorphisms.

Proof. We do this for one $\mathscr{E} = \mathscr{E}_1$, noting that the integers m_i which arise can be chosen uniformly for any finite collection of $\mathscr{E}_i$'s. By the preceding lemma, take a large n_1 and a section s_1 of $\mathscr{E}(n_1)$ that is not zero at any $x \in S$. Define $\mathscr{G}_1$ by

$$\mathscr{O}_X \xrightarrow{s_1} \mathscr{E}_1(n_1) \longrightarrow \mathscr{G}_1 \longrightarrow 0.$$

By Lemma 3.3 there is, if $r > 1$, for large n_2 a section s_2 of $\mathscr{E}_1(n_1 + n_2)$ whose image in $\mathscr{G}_1(n_2)$ is not zero at any $x \in S$. Define $\mathscr{G}_2$ by

$$\mathscr{L}^{\otimes n_2} \oplus \mathscr{O}_X \xrightarrow{s_1 \oplus s_2} \mathscr{E}(n_1 + n_2) \longrightarrow \mathscr{G}_2 \longrightarrow 0,$$

and take, if $r > 2$, s_3 in $\mathscr{E}(n_1 + n_2 + n_3)$ whose image in $\mathscr{G}_2(n_3)$ is not zero at any $x \in S$. Continuing in this way one arrives at

$$\mathscr{L}^{\otimes(n_2 + \cdots + n_r)} \oplus \cdots \oplus \mathscr{L}^{\otimes n_r} \oplus \mathscr{O}_X \xrightarrow{s_1 \oplus \cdots \oplus s_r} \mathscr{E}(n_1 + \cdots + n_r)$$

such that the induced map on fibres at $x \in S$ has rank r, so is an isomorphism. Tensoring this by $\mathscr{L}^{\otimes m}$, $m = -\sum n_i$, yields (3.2).

Proposition 3.5. *For any X with an ample invertible sheaf,*

$$F^1 K(X) = F^1_{\text{top}} K(X).$$

Proof. If $x \in F^1_{\text{top}} K(X)$, taking $Y = X$, we see that x is represented by a complex $\mathscr{E}^{\cdot}$ which is generically exact. Therefore

$$\varepsilon(x) = \sum (-1)^i \operatorname{rank}(\mathscr{E}^i) = 0,$$

so $x \in F^1 K(X)$.

Conversely if $x \in F^1 K(X)$, write $x = [\mathscr{E}_1] - [\mathscr{E}_2]$, with $\mathscr{E}_1$, $\mathscr{E}_2$ locally free of the same rank r. Given closed subsets Y_α of X, let S be the set of generic points of the irreducible components of the Y_α. Construct homomorphisms as in (3.2) of Lemma 3.4, for $\mathscr{E}_1$ and $\mathscr{E}_2$; each defines a complex $\mathscr{E}_i^{\cdot}$ with non-zero terms in degrees -1 and 0, whose support meets each Y_α in codimension at least one. Then

$$\mathscr{E}_1^{\cdot} \oplus \mathscr{E}_2^{\cdot}[-1]$$

represents x, with respect to $\{Y_\alpha\}$ and $n = 1$, so $x \in F^1_{\text{top}} K(X)$, as required.

Lemma 3.6. *Let $\mathscr{E}$ be locally free on X of rank $r + 1$, $f\colon \mathbf{P}(\mathscr{E}) \to X$ the associated projective bundle. Then there is an element z in $F^r_{\text{top}} K(\mathbf{P}(\mathscr{E}))$ such that*

$$f_K(z) = 1 \qquad \textit{in } K(X).$$

Proof. By Proposition 3.5, if $\ell = [\mathscr{O}_{\mathbf{P}}(1)]$, $1 - \ell^{-1}$ is in $F^1_{\text{top}} \mathbf{P}$, so

$$z = (1 - \ell^{-1})^r \in F^r_{\text{top}} \mathbf{P}.$$

Since $f_K(1) = 1$, and $f_K(\ell^{-i}) = 0$ for $1 \leqq i \leqq r$ (Corollary 2.4), the lemma follows.

Let $f: \mathbf{P} \to X$ be a projective bundle. We shall say that a (bounded) complex $\mathscr{E}^{\cdot}$ of locally free sheaves on $\mathbf{P}$ is **regular** if each $\mathscr{E}^i$, each $\operatorname{Ker}(d^i)$, each $\operatorname{Im}(d^i)$, and each homology sheaf $\mathscr{H}^i(\mathscr{E}^{\cdot})$ is regular in the sense of §2. Since the push-forward of a short exact sequence of regular sheaves is exact, and each $f_*\mathscr{E}^i$ is locally free, it follows that $f_*(\mathscr{E}^{\cdot})$ is a bounded complex of locally free sheaves on X, with

$$\mathscr{H}^i(f_*\mathscr{E}^{\cdot}) = f_*\mathscr{H}^i(\mathscr{E}^{\cdot}).$$

If a regular complex $\mathscr{E}^{\cdot}$ represents an element $x \in K(\mathbf{P})$, it follows from the definition of the push-forward f_K that $f_*\mathscr{E}^{\cdot}$ represents $f_K(x)$.

Lemma 3.7. *Let $x \in K(\mathbf{P})$ be represented by a complex $\mathscr{E}^{\cdot}$ which is exact on an open subset U of $\mathbf{P}$. Then there is a regular complex $\tilde{\mathscr{E}}$, exact on U, which represents x.*

Proof. As in Proposition 2.1, using the canonical exact sequences (2.3), it follows that for any locally free sheaf $\mathscr{F}$ on $\mathbf{P}$ and any n_0, there are locally free sheaves $\mathscr{F}^0, \ldots, \mathscr{F}^N$ on $\mathbf{P}$ with

$$(*) \qquad [\mathscr{F}] = \sum (-1)^i [\mathscr{F}^i] \qquad \text{in } K(\mathbf{P}),$$

$\mathscr{F}^i \cong f^*\mathscr{G}^i \otimes \mathscr{O}(m_i)$, $\mathscr{G}^i$ locally free on X, and $m_i \geqq n_0$. Indeed, (2.3) gives such $\mathscr{F}^0, \ldots, \mathscr{F}^{r+1}$ for $\mathscr{F}$ and $n_0 = 1$; given $(*)$ for some $\mathscr{F}$ and n_0, applying (2.3) to each $\mathscr{F}^i$ gives $(*)$ for $\mathscr{F}$ and $n_0 + 1$.

Now given $\mathscr{E}^{\cdot}$ representing x, choose n_0 so that for each of the sheaves $\mathscr{A} = \mathscr{E}^i$, $\operatorname{Ker}(d^i)$, $\operatorname{Im}(d^i)$, and $\mathscr{H}^i(\mathscr{E}^{\cdot})$, the sheaf $\mathscr{A}(n_0)$ is regular. Then choose $\mathscr{F}^0, \ldots, \mathscr{F}^N$ so that $(*)$ holds for $\mathscr{F} = \mathscr{O}_{\mathbf{P}}$, and this n_0. By the projection formula **R 3** of §2, each of the complexes $\mathscr{E}^{\cdot} \otimes \mathscr{F}^i$ is regular; by $(*)$, x is represented by the regular complex

$$\bigoplus_{i=0}^{N} \mathscr{E}^{\cdot} \otimes \mathscr{F}^i[i]$$

which is exact wherever $\mathscr{E}^{\cdot}$ is exact; as in Proposition 3.2, $[i]$ denotes a shift of the preceding complex.

Remark. Using the same canonical resolutions (2.3), one sees in fact that any complex $\mathscr{E}^{\cdot}$ on $\mathbf{P}$ admits a homomorphism

$$\mathscr{E}^{\cdot} \to \tilde{\mathscr{E}}^{\cdot}$$

to a regular complex $\tilde{\mathscr{E}}^{\cdot}$ which induces an isomorphism on homology sheaves.

Lemma 3.8. *Let $f: \mathbf{P} \to X$ be a projective bundle, and let $x \in K(X)$. If $f^K(x)$ is in $F^n_{\text{top}}K(P)$, then x is in $F^n_{\text{top}}K(X)$.*

Proof. Choose $z \in F^r_{\text{top}}K(\mathbf{P})$ satisfying the conditions of Lemma 3.6. Then $f^K(x) \cdot z$ is in $F^{n+r}_{\text{top}}K(\mathbf{P})$, with

$$f_K(f^K(x) \cdot z) = x \cdot f_K(z) = x.$$

Let $\{Y_\alpha\}$ be a finite set of closed subsets of X. By Lemma 3.7 we may represent $f^K(x) \cdot z$ by a regular complex $\mathscr{E}^{\cdot}$ such that

$$\operatorname{codim}(|\mathscr{E}^{\cdot}| \cap f^{-1}(Y_\alpha), f^{-1}(Y_\alpha)) \geqq n + r$$

for all α. Then $f_* \mathscr{E}^{\cdot}$ is a complex representing x, with

$$|f_* \mathscr{E}^{\cdot}| = f(|\mathscr{E}^{\cdot}|) \qquad \text{and} \qquad \operatorname{codim}(|f_* \mathscr{E}^{\cdot}| \cap Y_\alpha, Y_\alpha) \geqq n.$$

(The last inequality follows from the fact that for any closed $Y \subset X$, and $Z \subset f^{-1}(Y)$,

$$\operatorname{codim}(f(Z), Y) \geqq \operatorname{codim}(Z, f^{-1}Y) - r.$$

Indeed, letting A be the local ring of Y at an irreducible component of $f(Z)$, this follows from the fact that

$$\dim A[T_1, \ldots, T_r] = \dim(A) + r$$

(cf. [Mat], 14.A).)

Theorem 3.9. *For any X with an ample invertible sheaf, and all n,*

$$F^n K(X) \subset F^n_{\text{top}} K(X).$$

Proof. Given $x \in F^n K(X)$, we may assume

$$x = \gamma^{k_1}(x_1) \cdot \cdots \cdot \gamma^{k_m}(x_m)$$

with $x_i \in F^1 K(X)$, $\sum k_i \geqq n$. By the splitting principle there is a morphism

$$f: X' \to X$$

which is a composite of projective bundle projections, such that each $f^K(x_i)$ can be written as a sum of differences $u - v$ of classes of invertible sheaves. For such line elements u, $v \in K(X')$,

$$\gamma_t(u - v) = \gamma_t(u - 1)/\gamma_t(v - 1) = (1 + (u - 1)t)/(1 + (v - 1)t).$$

Therefore

$$\gamma^k(u - v) = (-1)^{k-1}(u - v)(v - 1)^{k-1}.$$

Now by Proposition 3.5, $u - v$ and $v - 1$ are in $F^1_{\text{top}}K(X')$. Since $F^{\cdot}_{\text{top}}$ is a ring filtration, it follows that $f^K(x)$ is in $F^n_{\text{top}}K(X')$. By Lemma 3.8, x must be in $F^n_{\text{top}}K(X)$, as required.

Remark. In [SGA 6] a filtration $K(X)_n$ was defined by the condition that an element x is in $K(X)_n$ if for any one closed $Y \subset X$, x is represented by a complex $\mathscr{E}^{\cdot}$ whose support meets Y in codimension at least n. Hence clearly

$$F^n_{\text{top}}K(X) \subset K(X)_n.$$

In particular, Theorem 3.9 answers a question left open in [SGA 6], IV, 6.10: the γ-filtration $F^nK(X)$ is finer than the topological filtration $K(X)_n$.

All the statements and proofs of this section work equally well for the filtration $K(X)_n$. We prefer the filtration F^n_{top} because it is functorial:

If $f\colon Y \to X$ is a morphism, then

$$f^K(F^n_{\text{top}}K(X)) \subset F^n_{\text{top}}K(Y).$$

To see this, let $x \in F^n_{\text{top}}K(X)$, and let $\{Y_\alpha\}$ be a collection of closed subsets of Y. To show that f^Kx is represented by a complex with respect to $\{Y_\alpha\}$ and n, we may assume each Y_α is irreducible, by replacing each Y_α by all its irreducible components. Then stratify X by locally closed subsets X_β so that each

$$Y_\alpha \cap f^{-1}X_\beta \to X_\beta$$

is equidimensional (e.g. flat). Then if $\mathscr{E}^{\cdot}$ represents x with respect to $\{X_\beta\}$ and n, $f^*\mathscr{E}^{\cdot}$ represents f^Kx with respect to $\{Y_\alpha\}$ and n.

Corollary 3.10. *If* $\dim X \leqq d$, *then* $F^{d+1}K(X) = 0$.

Proof. Proposition 3.2(b) with Theorem 3.9.

Corollary 3.11. *The Chern character* ch *induces an isomorphism of* $\mathbf{Q}K(X)$ *with* $\mathbf{Q}\text{Gr}\,K(X)$.

Proof. Corollary 3.10 and Chapter III, Theorem 3.5.

Remark 1. The first Chern class c_1 determines an isomorphism

$$c_1 \colon \operatorname{Pic}(X) \to F^1K(X)/F^2K(X) = \operatorname{Gr}^1 K(X),$$

where $\operatorname{Pic}(X)$ is the multiplicative group of isomorphisms classes of invertible sheaves. This comes from Theorem 1.7 of Chapter III, giving an isomorphism of $\mathbf{L}$ with $\operatorname{Gr}^1 K(X)$, and the isomorphism of $\operatorname{Pic}(X)$ with $\mathbf{L}$ as observed in §1.

Remark 2. More information relating the Grothendieck filtration with geometric filtrations will be given in Chapter VI, §5.

V §4. Resolutions and Regular Imbeddings

In this section we assume that all schemes X under consideration satisfy the following axiom:

(∗) *Any coherent sheaf of $\mathcal{O}_X$-modules is the image of a locally free sheaf.*

Any scheme with an ample invertible sheaf, e.g. any scheme quasi-projective over an affine scheme, satisfies this axiom (cf. [H], III), which suffices for most applications. In fact, any scheme which is quasi-projective over a divisorial (e.g. a locally factorial or regular) base scheme satisfies (∗) ([B]).

We let:

$\mathfrak{S}_X$ = category of coherent sheaves on X which admit a finite locally free resolution, i.e. there exists a finite resolution

$$0 \to \mathscr{E}_n \to \mathscr{E}_{n-1} \to \cdots \to \mathscr{E}_1 \to \mathscr{E}_0 \to \mathscr{F} \to 0 \tag{4.1}$$

with $\mathscr{E}_i$ locally free for all i.

Since locally free sheaves are in $\mathfrak{S}_X$, there is a canonical homomorphism

$$K(X) \to K(\mathfrak{S}_X).$$

Proposition 4.1. *This homomorphism is an isomorphism*

$$K(X) \cong K(\mathfrak{S}_X).$$

Its inverse is given by mapping a class $[\mathscr{F}]$ *on the alternating sum*

$$[\mathscr{F}] \mapsto \sum_{i \geqq 0} (-1)^i [\mathscr{E}_i],$$

where $\mathscr{E}.$ *is a finite locally free resolution of* $\mathscr{F}$.

Proof. It is a standard lemma in the theory of Grothendieck groups (cf. [L], IV, 3.7) that the above alternating sum gives a well-defined inverse isomorphism, because the following property is satisfied: if $\mathscr{F}$ admits a finite resolution of length n by locally free sheaves $\mathscr{E}_i$ and

$$0 \to \mathscr{H} \to \mathscr{E}'_{n-1} \to \cdots \to \mathscr{E}'_0 \to \mathscr{F} \to 0$$

is an exact sequence with $\mathscr{E}'_i$ a locally free sheaf for $0 \leqq i < n$, then $\mathscr{H}$ is also locally free; cf. the remarks in the introduction to this chapter.

If $f: X \to Y$ is a (closed) regular imbedding, then Koszul complexes (Chapter IV, §3) give, locally on Y, a resolution of $f_*\mathscr{O}_X$ of length r by locally free sheaves on Y, where r is the codimension of X in Y. Let $\mathscr{F}$ be locally free on X. By (*) we can find an exact sequence

$$\mathscr{E}_{r-1} \to \cdots \to \mathscr{E}_0 \to f_*\mathscr{F} \to 0$$

with $\mathscr{E}_i$ locally free on Y for $i = 0, \ldots, r-1$. Let $\mathscr{E}_r$ be the kernel of the arrow furthest to the left. By the above remarks, it follows that $\mathscr{E}_r$ is also locally free, and we obtain a locally free resolution

$$0 \to \mathscr{E}_r \to \mathscr{E}_{r-1} \to \cdots \to \mathscr{E}_0 \to f_*\mathscr{F} \to 0.$$

Thus we have shown that for any locally free sheaf $\mathscr{F}$ on X the direct image $f_*\mathscr{F}$ admits a finite resolution by locally free sheaves on Y. Since $f_*\mathscr{F}$ is just the extension of $\mathscr{F}$ by zero outside X, the functor f_* is an exact functor from the category $\mathfrak{V}_X$ of locally free sheaves on X to the category $\mathfrak{S}_Y$ of finitely resolvable sheaves on Y. This induces a homomorphism

$$f_K: K(X) \to K(Y).$$

Explicitly, if $0 \to \mathscr{E}_n \to \cdots \to \mathscr{E}_0 \to f_*\mathscr{F} \to 0$ is a resolution of $f_*\mathscr{F}$ by locally free sheaves $\mathscr{E}_i$ on Y, then

$$f_K([\mathscr{F}]) = \sum_{i=0}^{n} (-1)^i [\mathscr{E}_i].$$

Lemma 4.2 (Projection Formula). *If $f: X \to Y$ is a regular imbedding, then*

$$f_K(x \cdot f^K y) = f_K(x) \cdot y$$

for $x \in K(X)$, $y \in K(Y)$.

Proof. Let $x = [\mathscr{F}]$, $y = [\mathscr{G}]$, with $\mathscr{F}$, $\mathscr{G}$ locally free sheaves on X and Y. If $\mathscr{E}.$ is a resolution of $f_*\mathscr{F}$, then $\mathscr{G} \otimes \mathscr{E}.$ is a resolution of

$$f_*(f^*\mathscr{G} \otimes \mathscr{F}) = \mathscr{G} \otimes f_*\mathscr{F},$$

from which the formula follows.

Proposition 4.3. *Let $\mathscr{E}$ be a locally free sheaf of rank d on a scheme Y, s a regular section of $\mathscr{E}$, $X = Z(s)$ the zero scheme of s, f the imbedding of X in Y. Then f is a regular imbedding of codimension d. Let $e = [\mathscr{E}]$ in $K(Y)$. Then:*

(a) $$f_K(1) = \lambda_{-1}(e^\vee) \text{ in } K(Y);$$

(b) $$f_{\mathrm{Gr}\,K}(1) = c^{\mathrm{top}}(e) \quad \text{in } \mathrm{Gr}\, K(Y).$$

Proof. We have seen that f is regular, and that there is an exact Koszul resolution

$$0 \to \Lambda^d\mathscr{E}^\vee \to \Lambda^{d-1}\mathscr{E}^\vee \to \cdots \to \Lambda^1\mathscr{E}^\vee \to \mathscr{O}_Y \to f_*\mathscr{O}_X \to 0$$

as in Chapter IV, §2. This proves (a), and (b) then follows from Chapter III, Proposition 2.1.

To make use of the deformation to the normal bundle, we shall also need the following two propositions.

Proposition 4.4. *Let A, B, C be effective Cartier divisors on a scheme M. Assume:*

(i) $\mathscr{O}(A) \cong \mathscr{O}(B + C)$;
(ii) *B and C meet regularly in M.*

Let $D = B \cap C$, and let a, b, c, d be the imbeddings of A, B, C, D in M. Then

(a) $$a_K(1) = b_K(1) + c_K(1) - d_K(1) \text{ in } K(M);$$

(b) $$a_{\mathrm{Gr}\,K}(1) = b_{\mathrm{Gr}\,K}(1) + c_{\mathrm{Gr}\,K}(1) \qquad \text{in } \mathrm{Gr}^1\, K(M).$$

Proof. By (ii), D is the zero-scheme of a regular section of

$$\mathcal{O}(B) \oplus \mathcal{O}(C).$$

Using the preceding proposition and (i), we get

$$\begin{aligned} d_K(1) &= \lambda_{-1}(\mathcal{O}(B) \oplus \mathcal{O}(C)) \\ &= 1 - [\mathcal{O}(-B) \oplus \mathcal{O}(-C)] + [\mathcal{O}(-B-C)] \\ &= 1 - [\mathcal{O}(-B)] + 1 - [\mathcal{O}(-C)] - (1 - [\mathcal{O}(-A)]) \\ &= b_K(1) + c_K(1) - a_K(1). \end{aligned}$$

This proves (a). Formula (b) follows from (i) and Proposition 4.3, since

$$c_1(\mathcal{O}(A)) = c_1(\mathcal{O}(B + C)) = c_1(\mathcal{O}(B)) + c_1(\mathcal{O}(C)).$$

Proposition 4.5. *Let $F: P \to M$ be a regular embedding, and let*

$$\varphi: Y \to M$$

be a morphism. Form the fiber square:

$$\begin{array}{ccc} X & \xrightarrow{f} & Y \\ \psi \downarrow & & \downarrow \varphi \\ P & \xrightarrow[F]{} & M \end{array}$$

Assume that f is a regular imbedding of the same codimension as F. (This is true for instance if φ is a regular imbedding and P, Y meet regularly in M, in which case $X = P \cap Y$.) Then:

(a) $$\varphi^K F_K = f_K \psi^K.$$

If Z is a subscheme of Y which is disjoint from $f(X)$, and $h: Z \to M$ is the morphism induced by φ, then

(b) $$h^K F_K = 0.$$

Proof. Let $p \in K(P)$. We may assume that $p = [\mathscr{F}]$ with some locally free sheaf $\mathscr{F}$ on P. Let

$$0 \to \mathscr{E}_n \to \cdots \to \mathscr{E}_0 \to F_*(\mathscr{F}) \to 0$$

be a locally free resolution. We have $\varphi^* F_*(\mathscr{F}) = f_* \psi^*(\mathscr{F})$ since F_* and f_* are the extension by 0 because F, f are closed imbeddings. Hence to prove the proposition, it suffices to show that the sequence

$$\varphi^* \mathscr{E}. \to \varphi^* F_*(\mathscr{F}) \to 0$$

is exact. Taking φ^* locally amounts to taking the tensor product, and by abstract nonsense of basic homological algebra, the homology of the complex

$$0 \to \varphi^* \mathscr{E}_n \to \cdots \to \varphi^* \mathscr{E}_0 \to \varphi^* F_*(\mathscr{F}) \to 0$$

is independent of the choice of locally free resolution of $F_*(\mathscr{F})$. Hence the desired assertion is local on M, and we may assume that $M = \operatorname{Spec}(A)$ and X is defined by an ideal

$$I = (a_1, \ldots, a_r) = (a),$$

where (a) is a regular sequence. Also since $\mathscr{F}$ is locally free, we may assume that $\mathscr{F} = \mathscr{O}_P$, and $\mathscr{E}$ is the Koszul complex

$$0 \to K_r(a) \to \cdots \to K_0(a) \to A/I \to 0.$$

Taking φ^* amounts to tensoring with the structure sheaf of Y, and locally on Y we obtain the Koszul complex

$$0 \to K_r(\bar{a}) \to \cdots \to K_0(\bar{a}) \to B/\bar{I} \to 0,$$

where, say, $Y = \operatorname{Spec}(B)$, $\bar{I} = IB = A/I \otimes B$, and $\bar{a}_i$ is the natural image of a_i in B. But the assumption that $f: X \to Y$ is a regular imbedding of the same codimension as F implies that $(\bar{a})$ is a minimal set of generators for the ideal of X in Y locally. By Lemma 2.6 of Chapter IV it follows that $(\bar{a})$ is a regular sequence, thereby proving (a) of the proposition.

Assertion (b) follows from (a), or from the observation that, with $\mathscr{E}.$ as above, $h^* \mathscr{E}.$ is an exact complex on Z. This concludes the proof.

Remark. Proposition 4.5 will be substantially generalized later in Chapter VI, Proposition 1.1 and Theorem 1.3.

Classically, geometers work with an intersection product of classes of cycles on a scheme (or variety). The K-groups can be viewed as a substitute for cycle classes, and the product in $K(X)$ can be viewed as a substitute for the intersection product. In SGA 6 and [Man], global intersection formulas are proved using resolutions and Tor (see also [L], Chapter XVI, Theorem 10.11 and Proposition 11.1), after Serre's local theory (Springer Lecture Notes 11, 1965). Here we shall give such a formula as an application of Proposition 4.5, illustrating the special case already mentioned in its statement.

Corollary 4.6. *Let Y, Z be closed subschemes of X, regularly imbedded and meeting regularly in X. Then*

$$[\mathcal{O}_Y][\mathcal{O}_Z] = [\mathcal{O}_{Y \cap Z}].$$

Proof. Let $i: Y \to X$ and $j: Z \to X$ be the regular imbeddings of subschemes, and form the fibre square as shown:

$$\begin{array}{ccc} Y \cap Z & \xrightarrow{g} & Z \\ {\scriptstyle h}\downarrow & & \downarrow{\scriptstyle j} \\ Y & \xrightarrow[i]{} & X \end{array}$$

Then we have:

$$\begin{aligned}
[\mathcal{O}_Y][\mathcal{O}_Z] &= i_K(1)j_K(1) & \\
&= j_K j^K(i_K(1)) & \text{by projection formula} \\
&= j_K(g_K h^K(1)) & \text{by Proposition 4.5} \\
&= (j \circ g)_K(1) & \text{because } h^K(1) = 1 \\
&= [\mathcal{O}_{Y \cap Z}]. &
\end{aligned}$$

This proves the corollary.

The rest of this section is devoted to the proof of two lemmas, which are needed to construct a more general push-forward map in the next section, and to verify compatibilities of push-forward homomorphisms for imbeddings and projections.

Lemma 4.7. *Let $\mathscr{E}$ be a locally free sheaf of rank $r+1$ on X with associated projective bundle*

$$f: \mathbf{P}(\mathscr{E}) \to X.$$

Let $s: X \to \mathbf{P}(\mathscr{E})$ be a section of f, i.e., $f \circ s = \mathrm{id}_X$. Then s is a regular imbedding of codimension r, and

$$f_K \circ s_K = \mathrm{id}_{K(X)}.$$

Proof. We saw in Chapter IV, Lemma 3.8 that any section of a smooth morphism is regular, the present case of a bundle being particularly simple. To prove the last assertion, let $\mathscr{F}$ be a locally free sheaf on X. Then $s_*\mathscr{F}$ is a regular coherent sheaf on $\mathbf{P}(\mathscr{E})$. To see this, note that

$$s_*(\mathscr{F}) \otimes \mathscr{O}(k) = s_*(\mathscr{F} \otimes s^*\mathscr{O}(k)).$$

Since the restriction of f to $s(X)$ is an isomorphism,

$$R^i f_*(s_*(\mathscr{F} \otimes s^*\mathscr{O}(k))) = \begin{cases} 0, & i > 0, \\ \mathscr{F} \otimes s^*\mathscr{O}(k), & i = 0, \end{cases}$$

from which it follows that $s_*\mathscr{F}$ is regular.

By the construction of §2, $s_*\mathscr{F}$ has a canonical resolution

$$0 \to f^*\mathscr{T}_r(-r) \to \cdots \to f^*\mathscr{T}_0 \to s_*\mathscr{F} \to 0.$$

We claim the each $\mathscr{T}_p = \mathscr{T}_p(s_*\mathscr{F})$ is a locally free sheaf on X. We prove this by induction on p, together with the assertion that if $\mathscr{Z}_p$ is the sheaf defined by (2.7), then

$$f_*(\mathscr{Z}_p(m)) \quad \text{is locally free for} \quad m > p.$$

Since $\mathscr{Z}_p(m)$ is regular for $m > p$, (2.7) determines exact sequences

$$0 \to f_*(\mathscr{Z}_p(p+i)) \to f_*(f^*\mathscr{T}_p(i)) \to f_*(\mathscr{Z}_{p-1}(p+i)) \to 0 \tag{4.2}$$

for $i > 0$. Note to start that $\mathscr{T}_0 = f_* s_*(\mathscr{F}) = \mathscr{F}$ is locally free, and

$$0 \to f_*(\mathscr{Z}_0(i)) \to f_*(f^*\mathscr{F}(i)) \to f_*(s_*\mathscr{F}(i)) \to 0.$$

Since $f_* f^* \mathscr{F}(i) = \mathscr{F} \otimes \mathrm{Sym}^i \mathscr{E}$ and $f_* s_* \mathscr{F}(i) = \mathscr{F}(i) \otimes s^* \mathcal{O}(i)$ are locally free, so is the kernel, which completes the proof for $p = 0$. Assuming the result for $p - 1$, then

$$\mathscr{T}_p(s_* \mathscr{F}) = f_*(\mathscr{Z}_{p-1}(p))$$

is locally free by induction. And $f_* f^* \mathscr{T}_p(i) = \mathscr{T}_p \otimes \mathrm{Sym}^i \mathscr{E}$ is locally free, so $f_*(\mathscr{Z}_p(p+i))$ is locally free for $i > 0$ by (4.2).

From this canonical resolution we have

$$s_K[\mathscr{F}] = \sum_{p=0}^{r} (-1)^p [f^* \mathscr{T}_p] \ell^{-p}$$

with $\ell = [\mathcal{O}(1)]$, and $\mathscr{T}_p = \mathscr{T}_p(s_* \mathscr{F})$ locally free sheaves on X. By Theorem 2.3 and Chapter I, Proposition 2.2, $f_K(\ell^{-p}) = 0$ for $p = 1, \ldots, r$, and

$$f_K(s_K[\mathscr{F}]) = f_K[f^* \mathscr{T}_0] = [\mathscr{T}_0] = [\mathscr{F}],$$

which concludes the proof.

Lemma 4.8. *Let $f: \mathbf{P}(\mathscr{E}) \to Y$ be a projective bundle, and let $i: X \to Y$ be a regular imbedding. Form the fibre square*

$$\begin{array}{ccc} \mathbf{P}(i^*\mathscr{E}) & \xrightarrow{j} & \mathbf{P}(\mathscr{E}) \\ \downarrow g & & \downarrow f \\ X & \xrightarrow[i]{} & Y \end{array}$$

Then j is a regular imbedding, and

$$i_K \circ g_K = f_K \circ j_K.$$

Proof. The regularity of j follows from Chapter IV, Proposition 3.5. From the definitions we have immediately

$$j_K g^K(x) = f^K i_K(x) \tag{4.3}$$

for all $x \in K(X)$. Also, if $\ell = [\mathcal{O}(1)]$ is the canonical generator of $K(\mathbf{P}(\mathscr{E}))$, then

$$j^K(\ell) = [j^* \mathcal{O}(1)]$$

is the canonical generator of $\mathbf{P}(i^*\mathscr{E})$. By Theorem 2.3, $K(\mathbf{P}(i^*\mathscr{E}))$ is generated by elements $g^K(x)\cdot j^K(\ell^n)$ for $n \geqq 0$; to prove the formula of the lemma it suffices to see that both sides agree on such elements. Note also that

$$i^K f_K(\ell^n) = i^K[\mathrm{Sym}^n\, \mathscr{E}] = [\mathrm{Sym}^n\, i^*\mathscr{E}] = g_K j^K(\ell^n). \tag{4.4}$$

Using the projection formula together with (4.3) and (4.4) we have

$$\begin{aligned} f_K j_K(g^K(x)\cdot j^K(\ell^n)) &= f_K(j_K(g^K(x))\cdot \ell^n) \\ &= f_K(f^K i_K(x)\cdot \ell^n) = i_K(x)\cdot f_K(\ell^n) \\ &= i_K(x\cdot i^K f_K(\ell^n)) = i_K(x\cdot g_K(j^K\ell^n)) \\ &= i_K g_K(g^K(x)\cdot j^K(\ell)), \end{aligned}$$

as required.

V §5. The *K*-Functor of Regular Morphisms

All schemes considered in this section will be Noetherian, connected schemes satisfying the condition $(*)$ *of* §4.

Recall from Chapter IV, §3 that a regular morphism $f: X \to Y$ is one which can be factored into a regular imbedding and a projection from a projective bundle, $f = p \circ i$. The purpose of this section is to show how $X \mapsto K(X)$ is a λ-ring functor (as defined in Chapter II, §3) on the category of regular morphisms. The contravariant property is trivially satisfied, and we have to deal with the covariance and the projection formula. For a regular morphism as above, we shall define the push-forward

$$f_K: K(X) \to K(Y).$$

Let p_K and i_K be the homomorphisms defined in §2 and §4, and define

$$f_K = p_K \circ i_K.$$

Proposition 5.1.

(1) *The homomorphism* $p_K \circ i_K$ *is independent of the factorization of* f.

(2) *If* $f: X \to Y$ *and* $g: Y \to Z$ *are regular morphisms, then* $g \circ f$ *is a regular morphism, and* $(g \circ f)_K = g_K \circ f_K$.

Proof. In Chapter IV, Proposition 3.12 we proved that $g \circ f$ is a regular morphism. We now consider several cases.

Case 1. If $f: X \to Y$ and $g: Y \to Z$ are regular imbeddings, then $g \circ f$ is a regular imbedding, and

$$(g \circ f)_K = g_K \circ f_K.$$

That $g \circ f$ is a regular imbedding was seen in Chapter IV, Proposition 3.4. If $\mathscr{F}$ is a locally free sheaf on X, and $\mathscr{E}.$ is a resolution of $f_*\mathscr{F}$ by locally free sheaves on Y, construct a double complex $\mathscr{D}..$ of locally free sheaves on Z:

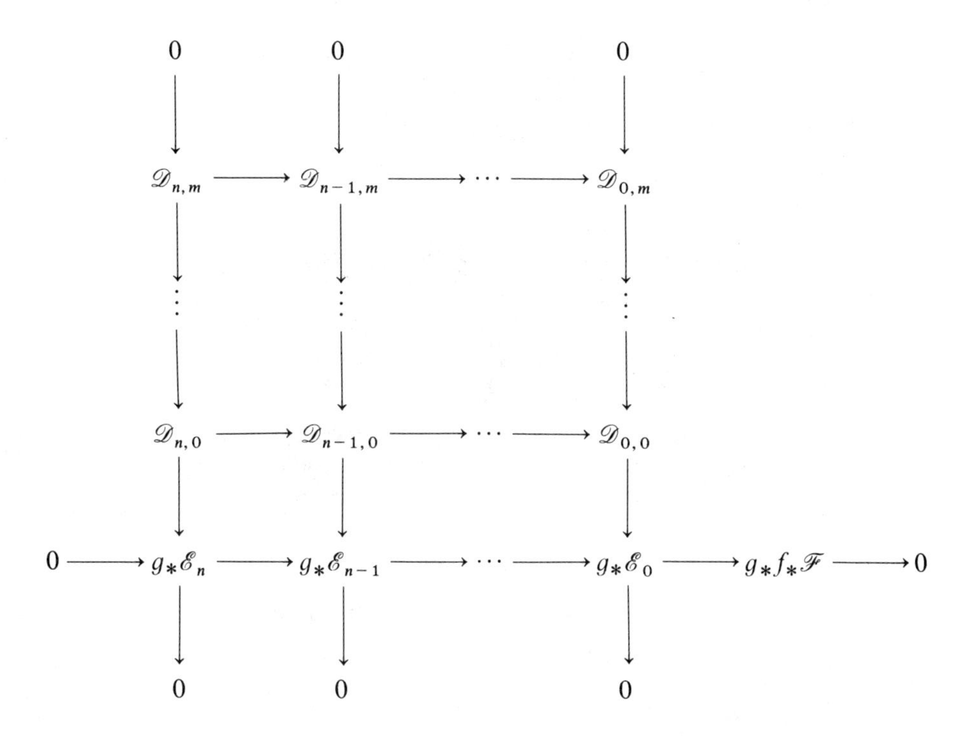

$$\begin{array}{ccccccccccc}
 & & 0 & & 0 & & & & 0 & & \\
 & & \downarrow & & \downarrow & & & & \downarrow & & \\
 & & \mathscr{D}_{n,m} & \to & \mathscr{D}_{n-1,m} & \to & \cdots & \to & \mathscr{D}_{0,m} & & \\
 & & \downarrow & & \downarrow & & & & \downarrow & & \\
 & & \vdots & & \vdots & & & & \vdots & & \\
 & & \downarrow & & \downarrow & & & & \downarrow & & \\
 & & \mathscr{D}_{n,0} & \to & \mathscr{D}_{n-1,0} & \to & \cdots & \to & \mathscr{D}_{0,0} & & \\
 & & \downarrow & & \downarrow & & & & \downarrow & & \\
0 & \to & g_*\mathscr{E}_n & \to & g_*\mathscr{E}_{n-1} & \to & \cdots & \to & g_*\mathscr{E}_0 & \to & g_*f_*\mathscr{F} \to 0 \\
 & & \downarrow & & \downarrow & & & & \downarrow & & \\
 & & 0 & & 0 & & & & 0 & &
\end{array}$$

so that the columns resolve the sheaves $g_*\mathscr{E}_i$. By Lemma 5.4 of the appendix to this section, applied to the homomorphism from $\mathscr{D}..$ to $g_*\mathscr{E}.$, the associated total complex of $\mathscr{D}..$ resolves $(gf)_*\mathscr{F}$, so

$$\begin{aligned}
(g \circ f)_K[\mathscr{F}] &= \sum_{i,j} (-1)^{i+j}[\mathscr{D}_{i,j}] \\
&= \sum_i (-1)^i \sum_j (-1)^j [\mathscr{D}_{i,j}] \\
&= \sum_i (-1)^i g_K[\mathscr{E}_i] \\
&= g_K(f_K[\mathscr{F}]).
\end{aligned}$$

Case 2. Let $\mathscr{E}$, $\mathscr{E}'$ be locally free sheaves on Y, $\mathbf{P} = \mathbf{P}(\mathscr{E})$, $\mathbf{P}' = \mathbf{P}(\mathscr{E}')$, p and p' the projections. Form the fibre square

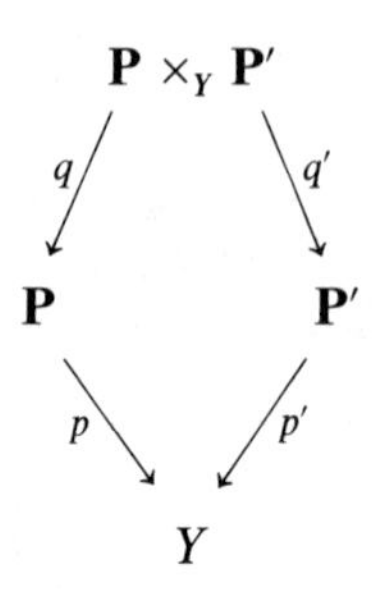

Then q and q' are projective bundle projections, and

$$p'_K \circ q'_K = p_K \circ q_K.$$

To see this, let ℓ and ℓ' be the canonical generators of $K(\mathbf{P})$ and $K(\mathbf{P}')$. Note that

$$q'_K q^K(\ell^a) = [p'^* \operatorname{Sym}^a \mathscr{E}] = p'^K p_K(\ell^a).$$

By Theorem 2.3 the classes $q^K(\ell^a) \cdot q'^K(\ell'^b)$ generate $K(\mathbf{P} \times_Y \mathbf{P}')$ over $K(Y)$, so it suffices to show that $p'_K \circ q'_K$ and $p_K \circ q_K$ agree on such classes. Using the preceding equation, with the projection formula,

$$\begin{aligned} p'_K \circ q'_K(q^K(\ell^a) q'^K(\ell'^b)) &= p'_K(q'_K(q^K(\ell^a)) \cdot \ell'^b) \\ &= p'_K(p'^K(p_K(\ell^a)) \cdot \ell'^b) = p_K(\ell^a) \cdot p'_K(\ell'^b). \end{aligned}$$

By symmetry, this equals $p_K \circ q_K(q^K(\ell^a) \cdot q'^K(\ell'^b))$, which concludes the proof in this case.

Case 3. Suppose $f: X \to Y$ is a regular imbedding which factors through a projective bundle

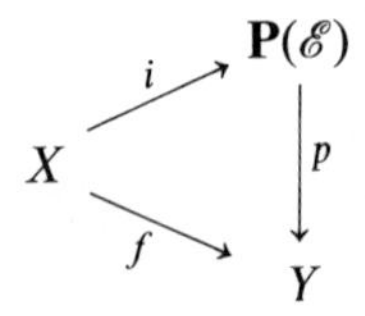

Then i is a regular imbedding, and

$$f_K = p_K \circ i_K.$$

For the proof, let $q: \mathbf{P}(f^*\mathscr{E}) \to X$ be the induced projective bundle, and let s be the section induced by i:

$$\begin{array}{ccc} \mathbf{P}(f^*\mathscr{E}) & \xrightarrow{\ j\ } & \mathbf{P}(\mathscr{E}) \\ q\downarrow\uparrow s & \nearrow^{i} & \downarrow p \\ X & \xrightarrow[f]{} & Y \end{array}$$

By Chapter IV, Propositions 3.5 and 3.9, j, s, and i are regular imbeddings. By Case 1,

$$i_K = j_K \circ s_K.$$

By Lemmas 4.7 and 4.8,

$$q_K \circ s_K = \mathrm{id}_{K(X)} \qquad \text{and} \qquad f_K \circ q_K = p_K \circ j_K.$$

Therefore,

$$p_K \circ i_K = p_K \circ j_K \circ s_K = f_K \circ q_K \circ s_K = f_K,$$

as asserted.

We can now prove (1) of the theorem. Let $f = p \circ i = p' \circ i'$ be two such factorizations of f through projective bundles $\mathbf{P}$ and $\mathbf{P}'$. Form the commutative diagram

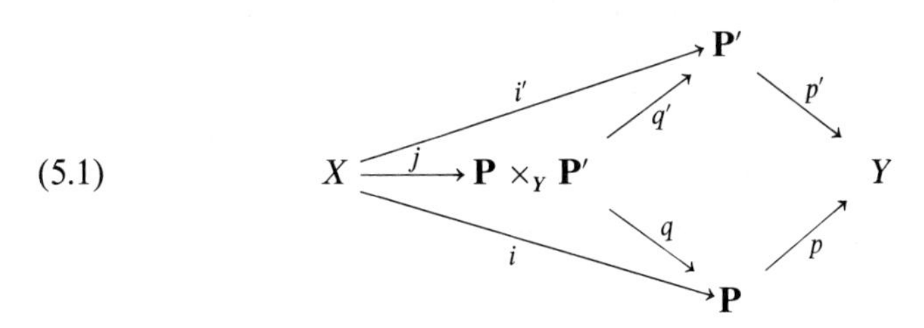

where $j = (i, i')$ is the diagonal imbedding. Since q and q' are projective bundle projections, Case 3 implies that j is a regular imbedding, with

$$i_K = q_K \circ j_K \qquad \text{and} \qquad i'_K = q'_K \circ j_K.$$

By Case 2,

$$p_K \circ i_K = p_K \circ q_K \circ j_K = p'_K \circ q'_K \circ j_K = p'_K \circ i'_K,$$

as required.

Next we prove the second part of the theorem. Let

$$X \xrightarrow{i} \mathbf{P}\mathscr{E} \xrightarrow{p} Z$$

be a factorization of $g \circ f$ into a closed imbedding followed by a bundle projection. This determines a commutative diagram:

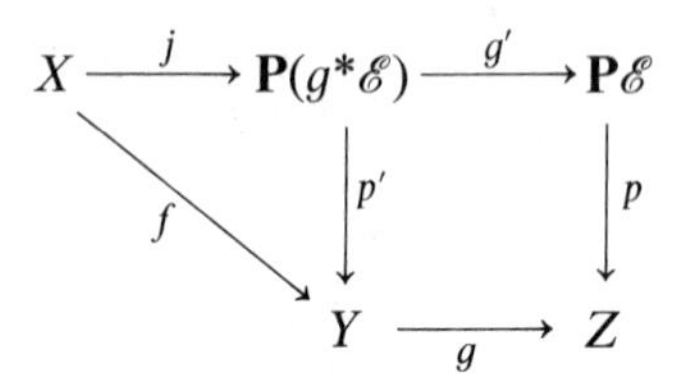

with $i = g' \circ j$. Then

$$\begin{aligned} g_K \circ f_K &= g_K \circ p'_K \circ j_K && \text{by (1)} \\ &= p_K \circ g'_K \circ j_K && \text{by Lemma 4.8} \\ &= p_K \circ i_K && \text{by Case 3} \\ &= (g \circ f)_K && \text{by definition,} \end{aligned}$$

which concludes the proof.

Remark. A more conceptual but less elementary proof of the proposition can be given along the following lines. If $f: X \to Y$ is a proper regular morphism, and $\mathscr{E}^{\cdot}$ is a bounded complex of locally free sheaves on X, one can show that the complex $Rf_*(\mathscr{E}^{\cdot})$ in the derived category is homologically isomorphic to a bounded complex $\mathscr{F}^{\cdot}$ of locally free sheaves on Y, and that the Euler characteristic

$$\sum (-1)^i [\mathscr{F}^i] \qquad \text{in } K(Y)$$

is independent of choice of $\mathscr{F}^{\cdot}$. Then

$$\sum (-1)^i f_K[\mathscr{E}^i] = \sum (-1)^i [\mathscr{F}^i].$$

This description is independent of factorization; the functoriality follows from the equation

$$R(g \circ f) = R(g) \circ R(f)$$

in the derived category. This approach also generalizes to "perfect" morphisms; for details, see [SGA 6].

Proposition 5.2 (Projection Formula). *For a regular morphism $f: X \to Y$, $x \in K(X)$, $y \in K(Y)$,*

$$f_K(x \cdot f^K y) = f_K(x) \cdot y.$$

Proof. This follows from Lemmas 2.6 and 4.2 which proved the projection formula for each one of the cases of a projection from a projective bundle and a regular imbedding respectively.

We can now summarize our results in the following theorem.

Theorem 5.3. *On the category of regular morphisms, $X \mapsto K(X)$ is a λ-ring functor.*

Remark. When Y has an ample invertible sheaf $\mathscr{L}$, a projective morphism $f: X \to Y$ admits a factorization into

$$X \xrightarrow{i} \mathbf{P}_Y^n \xrightarrow{p} Y,$$

i a closed imbedding, p the projection. To see this, factor f through $\mathbf{P}(\mathscr{E})$ as usual, and take n and m so there is a surjection

$$\mathscr{O}_Y^{\oplus(n+1)} \to \mathscr{E} \otimes \mathscr{L}^{\otimes m} \to 0.$$

This induces a closed imbedding

$$\mathbf{P}(\mathscr{E}) \cong \mathbf{P}(\mathscr{E} \otimes \mathscr{L}^{\otimes m}) \hookrightarrow \mathbf{P}(\mathscr{O}_Y^{\oplus(n+1)}) = \mathbf{P}_Y^n,$$

as required.

With this remark, it is only necessary to study trivial projections $\mathbf{P}_Y^n \to Y$. For several Riemann–Roch theorems, this simplifies the computations considerably.

Homological Appendix

We have used a basic lemma from homological algebra, which is usually proved using spectral sequences. For convenience of the reader, we include an elementary treatment here, following the general principle that double complexes can be used directly.

By a **double complex** in some abelian category we mean a commutative diagram

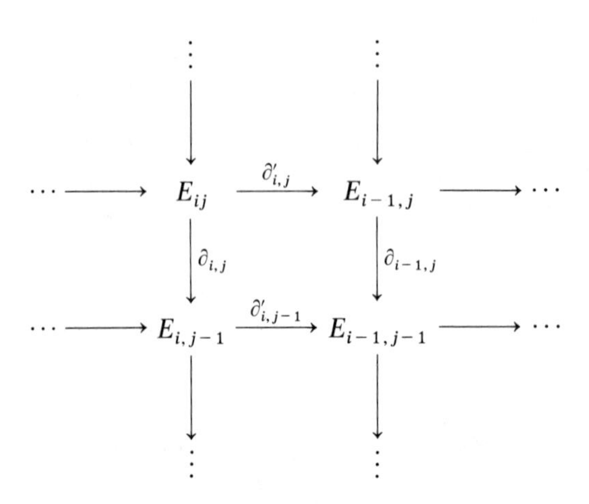

whose columns and rows are all complexes, denoted $E_{i\cdot}$ and $E_{\cdot j}$ respectively. We assume the complexes are bounded below, i.e. $E_{ij} = 0$ for $i < N$, $j < N$, some N. The **associated total complex** $\mathrm{Tot}(E_{\cdot\cdot})$ is the complex whose n-th term is

$$\mathrm{Tot}(E_{\cdot\cdot})_n = \bigoplus_{i+j=n} E_{ij}$$

and whose n-th boundary d_n is the sum of homomorphisms

$$(\partial_{ij}, (-1)^j \partial'_{ij}) \colon E_{ij} \to E_{i,j-1} \oplus E_{i-1,j}.$$

A homomorphism $\varphi_{\cdot\cdot} \colon E_{\cdot\cdot} \to F_{\cdot\cdot}$ of double complexes induces a homomorphism of complexes

$$\mathrm{Tot}(\varphi_{\cdot\cdot}) \colon \mathrm{Tot}(E_{\cdot\cdot}) \to \mathrm{Tot}(F_{\cdot\cdot}),$$

as well as homomorphisms of column complexes

$$\varphi_{i\cdot} \colon E_{i\cdot} \to F_{i\cdot},$$

and similarly for the rows.

Recall that a homomorphism $\varphi_{\cdot} \colon E_{\cdot} \to F_{\cdot}$ of complexes is called a **homology isomorphism** (or quasi-isomorphism) if the induced homomorphisms

$$H_i(\varphi_{\cdot}) \colon H_i(E_{\cdot}) \to H_i(F_{\cdot})$$

are all isomorphisms.

Lemma 5.4. *Let* $\varphi_{..}: E_{..} \to F_{..}$ *be a homomorphism of double complexes such that each homomorphism* $\varphi_{i.}$ *is a homology isomorphism. Then* $\mathrm{Tot}(\varphi_{..})$ *is a homology isomorphism.*

Proof. For a double complex $E_{..}$, let $E_{..}(r)$ denote the truncation of $E_{..}$ obtained by omitting all columns $E_{i.}$ of $E_{..}$ with $i > r$. Then $E_{..}(r)$ is a subcomplex of $E_{..}$, with quotient double complex denoted $\bar{E}_{..}(r)$. From a homomorphism $\varphi_{..}$ one has a commutative diagram

$$\begin{array}{ccccccccc} 0 & \longrightarrow & E_{..}(r) & \longrightarrow & E_{..} & \longrightarrow & \bar{E}_{..}(r) & \longrightarrow & 0 \\ & & \downarrow{\scriptstyle \varphi_{..}(r)} & & \downarrow{\scriptstyle \varphi_{..}} & & \downarrow{\scriptstyle \bar{\varphi}_{..}(r)} & & \\ 0 & \longrightarrow & F_{..}(r) & \longrightarrow & F_{..} & \longrightarrow & \bar{F}_{..}(r) & \longrightarrow & 0 \end{array}$$

of double complexes, with exact rows.

Since, for a given n, $\mathrm{Tot}(E_{..})$ and $\mathrm{Tot}(E_{..}(r))$ have the same n-th homology, for r sufficiently large, it suffices to prove the lemma in case $E_{..}$ and $F_{..}$ have only a finite number of non-zero columns. We prove this by induction on the number of columns i for which $E_{i.}$ or $F_{i.}$ is non-zero. If this number is one, the assertion is trivial, since the total complex is the same as the non-zero column. Otherwise one may choose an integer r so that the complexes $E_{..}(r)$ and $F_{..}(r)$, as well as the quotients $\bar{E}_{..}(r)$ and $\bar{F}_{..}(r)$, have fewer non-zero columns. By induction $\mathrm{Tot}(\varphi_{..}(r))$ and $\mathrm{Tot}(\bar{\varphi}_{..}(r))$ are homology isomorphisms. From the above diagram of double complexes one has a corresponding diagram of total complexes, also with exact rows. From the long exact homology sequences, and the Five Lemma, it follows that each $H_n(\mathrm{Tot}(\varphi_{..}))$ is an isomorphism, as required.

For the next two sections §6, §7, we work under the following conditions. We fix an affine Noetherian base scheme S. Let $\mathfrak{C}$ *be the category whose objects are connected schemes X which are quasi-projective over S, and whose morphisms are regular morphisms, i.e. projective local complete intersection morphisms. Any X in* $\mathfrak{C}$ *satisfies* (∗) *of* V, §4, *namely a coherent sheaf on X is the image of a locally free sheaf.*

V §6. Adams Riemann-Roch for Imbeddings

Under the stated conditions, by Theorem 5.3, and referring back to Chapter II, §3 we have the Riemann-Roch functors

$$(K, \psi^j, K) \quad \text{with integers} \quad j \geqq 0,$$

where ψ^j is the Adams character.

Lemma 6.1. *Let $f: X \to Y$ be a regular imbedding. Let $Y' = \mathbf{P}(\mathscr{C}_{X/Y} \oplus \mathscr{O}_X)$, and let f' be the zero section of X in Y'. Then the deformation to the normal bundle constructed in Chapter* IV, §5 *makes f' a basic deformation of f with respect to the Riemann–Roch functor (K, ψ^j, K) $(j \geqq 0)$, in the sense of Chapter* II, §1.

Proof. We have to verify the four **BD** properties. Property **BD 4** of the definition of a basic deformation is valid by construction; **BD 2** follows from Proposition 4.4(a). To prove **BD 1** and **BD 3**, given $x \in K(X)$, let

$$\tilde{x} = \mathrm{pr}^K(x) \in K(\mathbf{P}^1_X),$$

where $\mathrm{pr}: \mathbf{P}^1_X \to X$ is the projection. Let

$$y = F_K(\tilde{x}) \qquad \text{in } K(M).$$

Then **BD 1** and **BD 3** follow from Proposition 4.5. This proves the lemma.

Lemma 6.2. *Let $\mathscr{E}$ be a locally free sheaf on X and let*

$$f: X \to \mathbf{P}(\mathscr{E} \oplus \mathscr{O}_X)$$

be the zero section imbedding. Then f is an elementary imbedding with respect to the λ-ring functor K in the sense of Chapter II, §3. *Let $\mathscr{Q}$ be the universal hyperplane sheaf on $\mathbf{P}(\mathscr{E} \oplus \mathscr{O}_X)$ (Chapter* IV, §1) *and let $q = [\mathscr{Q}]$. Then*

$$f_K(1) = \lambda_{-1}(q) \qquad \textit{and} \qquad f^*(\mathscr{Q}) = \mathscr{E}.$$

Proof. By Proposition 2.7 of Chapter IV we know that X is the zero-scheme of a regular section of the locally free sheaf $\mathscr{Q}^\vee$. The first formula giving $f_K(1)$ follows from Proposition 4.3(a), and the second giving $f^*(\mathscr{Q})$ follows from Proposition 3.2(b) of Chapter IV. This concludes the proof.

Theorem 6.3. *If $f: X \to Y$ is a regular imbedding, then Riemann–Roch holds for f with respect to (K, ψ^j, K), with multiplier $\theta^j(\mathscr{C}_{X/Y})$. In other words, the diagram*

$$\begin{array}{ccc} K(X) & \xrightarrow{\theta^j(\mathscr{C}_{X/Y})\cdot\psi^j} & K(X) \\ \downarrow{f_K} & & \downarrow{f_K} \\ K(Y) & \xrightarrow{\psi^j} & K(Y) \end{array}$$

commutes.

Proof. Since f admits a basic deformation to an elementary imbedding, Theorem 1.3 of Chapter II tells us that it suffices to prove Riemann–Roch for the deformation f'. But Lemma 6.2 shows that the abstract conditions of Riemann–Roch in Chapter II, Theorem 3.1 are satisfied here, and an application of this previous theorem concludes the proof.

Application to the Graded Degree

In Chapter III, we related the Adams Riemann–Roch theorem with the graded degree of f_K. Using the results of Chapter III, we can now prove that if $f: X \to Y$ is a morphism in $\mathfrak{C}$, then

$$f_K: \mathbf{Q}K(X) \to \mathbf{Q}K(Y)$$

has a graded degree in the sense of Chapter III, §2, thus completing the last preparations for the Riemann–Roch theorems of the next section.

Proposition 6.4.

(a) *If $f: X \to Y$ is a regular imbedding of codimension d, then for all n,*

$$f_K(\mathbf{Q}F^n K(X)) \subset \mathbf{Q}F^{n+d}K(Y).$$

(b) *If $\mathscr{E}$ is a locally free sheaf of rank $r+1$ on a scheme Y, $X = \mathbf{P}(\mathscr{E})$, and $f: X \to Y$ is the projection, then for all n,*

$$f_K(F^n K(X)) \subset F^{n-r}K(Y).$$

Proof. (a) follows from the preceding Theorem 6.3, and the implication (2) ⇒ (1) of Chapter III, Theorem 4.1; (b) follows from Corollary 2.4 and Chapter III, Corollary 1.3.

Warning. Although part (b) shows that f_K has a graded degree on the filtration for K, part (a) gives this result only after tensoring with $\mathbf{Q}$. This is apparently essential, cf. [SGA 6], XIV. This implies that the Riemann–Roch theorem in K-theory will have denominators.

Proposition 6.5. *If $f: X \to Y$ is a regular morphism of codimension d, then*

$$f_K(\mathbf{Q}F^n K(X)) \subset \mathbf{Q}F^{n+d}K(Y) \qquad \text{for all } n \in \mathbf{Z}.$$

Proof. The proposition follows from Proposition 6.4 by factoring f into a closed imbedding followed by a projection.

From Proposition 6.5 we conclude that f_K induces homomorphisms

$$f_G: \mathbf{Q}\mathrm{Gr}^n K(X) \to \mathbf{Q}\mathrm{Gr}^{n+d} K(Y).$$

It is convenient here to put $G = \mathbf{Q}\mathrm{Gr}\, K$.

Theorem 6.6. *The association* $X \mapsto G(X) = \mathbf{Q}\mathrm{Gr}\, K(X)$ *is a covariant functor from our category* $\mathfrak{C}$ *to graded groups. Furthermore,* $(K, c, \mathbf{Q}\mathrm{Gr}\, K)$ *is a Chern class functor, and* $(K, \mathrm{ch}, \mathbf{Q}\mathrm{Gr}\, K)$ *is a Riemann–Roch functor in the sense of Chapter* II, §1.

Proof. Proposition 6.5 shows that all morphisms in our category have a graded degree in the sense of Chapter III, §2, and that our present situation fits the axiomatized considerations therein, including the statement of the present theorem. Of course, the nilpotency is guaranteed by the much stronger condition of Corollary 3.10, that for each X there is an integer d such that $F^{d+1}K(X) = 0$.

We are now in a position to repeat Lemma 6.1 for the graded functor.

Lemma 6.7. *Let* $f: X \to Y$ *be a regular imbedding. Let* $Y' = \mathbf{P}(\mathscr{C}_{X/Y} \oplus \mathscr{O}_X)$, *and let* f' *be the zero section of* X *in* Y'. *Then the deformation to the normal bundle constructed in Chapter* IV, §5 *makes* f' *a basic deformation of* f *with respect to the Riemann–Roch functor* $(K, \mathrm{ch}, \mathbf{Q}\mathrm{Gr}\, K)$ *in the sense of Chapter* II, §1.

Proof. Same as for Lemma 6.1, using Proposition 4.4(b) instead of 4.4(a).

V §7. The Riemann–Roch Theorems

We continue with the same category described before §6.

Let $f: X \to Y$ be a morphism in $\mathfrak{C}$ and let

$$X \xrightarrow{\ i\ } P \xrightarrow{\ p\ } Y$$

be a factoring of f into a regular imbedding i followed by a smooth morphism p. Define the **tangent element**

$$T_f = [i^*(\Omega^1_{P/Y})^\vee)] - [(\mathscr{C}_{X/P})^\vee] = [i^*\mathscr{T}_{P/Y}] - [\mathscr{N}_{X/P}],$$

where $\mathscr{T}_{P/Y}$ is the relative tangent sheaf and $\mathscr{N}_{X/P}$ the normal sheaf. Often T_f is called the **virtual tangent bundle** of f. But it is not a bundle, it is an element of the K-group $K(X)$. Also see Remark 1 below.

Proposition 7.1.

(i) *The element* T_f *in* $K(X)$ *is independent of the factorization of* f.
(ii) *If* g, f *are regular morphisms such that* $g \circ f$ *is defined, then*

$$T_{g \circ f} = f^K T_g + T_f.$$

Proof. Given another factorization

$$X \to P' \to Y,$$

form a diagonal diagram as in the proof of Theorem 5.1:

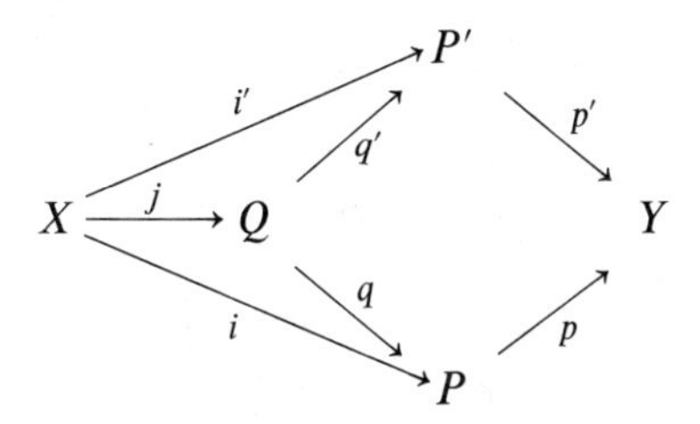

where $Q = P \times_Y P'$. By Chapter IV, Proposition 3.9 there is an exact sequence

$$0 \to \mathscr{C}_{X/P} \to \mathscr{C}_{X/Q} \to j^*\Omega^1_{Q/P} \to 0.$$

Since $\Omega^1_{Q/P} = q'^*\Omega^1_{P'/Y}$, this yields

$$[\mathscr{C}_{X/Q}] = [\mathscr{C}_{X/P}] - [i'^*\Omega^1_{P'/Y}].$$

By symmetry,

$$[\mathscr{C}_{X/Q}] = [\mathscr{C}_{X/P'}] - [i^*\Omega^1_{P/Y}].$$

Comparing these two equations and applying the involution $\vee$ gives the required equality in $K(X)$. This proves the first part of the proposition.

The second assertion of the proposition is an immediate consequence of the first assertion, together with the relations in the K-groups obtained from the short exact sequences of Chapter IV, Propositions 3.4, 3.7, 3.9. Each one of these exact sequences gives an additive relation of the desired type for the tangent element in special cases of composites, which when put together give the general relation as stated here.

Remark 1. If f is an imbedding then T_f is the negative of the class of the normal sheaf. If f is a smooth morphism, then T_f is the class of the relative tangent sheaf in the ordinary sense. Thus in general, T_f unifies these two notions in the K-group.

Remark 2. Since the Todd map is a homomorphism, the additivity of Proposition 7.1(ii) implies the multiplicativity of $\tau_f = \mathrm{td}(T_f)$ in towers, namely

$$\tau_{g \circ f} = f^G \tau_g \cdot \tau_f,$$

where $G = \mathbf{Q}\mathrm{Gr}\,\mathbf{K}$.

We now give for the Grothendieck Riemann–Roch theorem the statement corresponding to Lemma 6.2 for Adams Riemann–Roch.

Lemma 7.2. *Let $\mathscr{E}$ be a locally free sheaf on X and let*

$$f: X \to \mathbf{P}(\mathscr{E} \oplus \mathscr{O}_X)$$

be the zero section imbedding. Let $G = \mathbf{Q}\mathrm{Gr}\,K$. Then f is an elementary imbedding with respect to the Chern class functor (K, c, G) in the sense of Chapter II, §2. *If $\mathscr{Q}$ is the universal hyperplane sheaf on $\mathbf{P}(\mathscr{E} \oplus \mathscr{O}_X)$ and $q = [\mathscr{Q}]$, then*

$$f_K(1) = \lambda_{-1}(q) \qquad \text{and} \qquad f_G(1) = c^{\mathrm{top}}(q^\vee).$$

Proof. This is an immediate consequence of Proposition 4.3 and Lemma 6.2.

Theorem 7.3 (Grothendieck Riemann–Roch). *For any $f: X \to Y$ in $\mathfrak{C}$, Riemann–Roch holds for f with respect to* $(K, \mathrm{ch}, \mathbf{Q}\mathrm{Gr}\,K)$, *with multiplier*

$$\tau_f = \mathrm{td}(T_f).$$

In other words, the following diagram is commutative:

$$\begin{array}{ccc} K(X) & \xrightarrow{\mathrm{td}(T_f)\cdot \mathrm{ch}} & \mathbf{Q}\mathrm{Gr}\,K(X) \\ \downarrow{\scriptstyle f_K} & & \downarrow{\scriptstyle f_{\mathbf{Q}\mathrm{Gr}\,K}} \\ K(Y) & \xrightarrow[\mathrm{ch}]{} & \mathbf{Q}\mathrm{Gr}\,K(Y) \end{array}$$

Proof. Factor f into $p \circ i$, with i a regular imbedding and p a projective bundle projection. In Chapter II, Theorem 1.1 we showed that the Riemann–Roch theorem for two morphisms implies Riemann–Roch for their composite with a multiplier which is obtained precisely satisfying the formalism of the tangent element of Proposition 7.1(ii). Therefore it suffices to prove the Riemann-Roch theorem in the present context for a regular imbedding and a projection separately.

For an imbedding, we can use Lemma 7.2 and Lemma 6.7. They allow us to apply Theorem 2.1 of Chapter II, which says that Riemann-Roch is valid for elementary imbedding and Theorem 1.3 of Chapter II which says that if a morphism admits a basic deformation to an elementary imbedding, then Riemann-Roch holds for this morphism. Note that if f and f' are as in Lemma 6.7 or 7.2, then

$$\tau_f = \tau_{f'} = \mathrm{td}(\mathscr{C}^\vee_{X/Y})^{-1} = \mathrm{td}(f'^K(q^\vee))^{-1},$$

where $q = [\mathscr{Q}]$ and $\mathscr{Q}$ is the universal hyperplane sheaf on $\mathbf{P}(\mathscr{E} \oplus \mathscr{O}_X)$. This concludes the proof for regular imbeddings.

For the case of a projection $f\colon \mathbf{P}(\mathscr{E}) \to Y$, it follows from §2 that f is an elementary projection in the sense of Chapter II, §2, so Riemann-Roch holds with multiplier $\mathrm{td}(\ell e^\vee)$. By Chapter IV, Proposition 3.13, there is an exact sequence

$$0 \to \Omega^1_{\mathbf{P}/Y} \to f^*\mathscr{E} \otimes \mathscr{O}(-1) \to \mathscr{O}_{\mathbf{P}} \to 0,$$

so that

$$\mathrm{td}(\ell e^\vee) = \mathrm{td}((\Omega^1_{\mathbf{P}(\mathscr{E})/Y})^\vee),$$

as required. This concludes the proof of the Grothendieck Riemann-Roch theorem.

We make no attempt to list applications of Riemann-Roch here, but include the following famous special case. Suppose that X is a local complete intersection of dimension n over a field k. Let $Y = \mathrm{Spec}(k)$. Then

$$f_K\colon K(X) \to K(Y) = \mathbf{Z}$$

can be identified with the Euler characteristic $\chi_X = \chi(X, -)$, where

$$\chi(X, \mathscr{E}) = \sum_{i=0}^{n} (-1)^i \dim_k H^i(X, \mathscr{E}).$$

On the graded side,

$$f_G\colon \mathbf{Q}\mathrm{Gr}\, K(X) \to \mathbf{Q}\mathrm{Gr}\, K(Y) = \mathbf{Q}$$

is called the **top graded degree**, and then f_G is often denoted by $\int_X$. For a further description of f_G in this case, see Chapter VI, the example following Corollary 5.4. Finally the tangent element denoted by $[\mathcal{T}_X]$ is the class in the K-group of the tangent sheaf if X is smooth, otherwise is defined as we did previously using a regular embedding of X into a smooth variety, or into a projective space over k. Therefore the Grothendieck Riemann–Roch theorem implies:

Corollary 7.4 (Hirzebruch Riemann–Roch). *Let X be a local complete intersection of dimension n over a field k. Then for any locally free sheaf $\mathcal{E}$ on X, we have*

$$\chi(X, \mathcal{E}) = \int_X \mathrm{ch}(\mathcal{E})\,\mathrm{td}[\mathcal{T}_X].$$

As an application, if $[\mathcal{T}_X] = n$ where $n = \dim X$ (for instance if X is an abelian variety so $\mathcal{T}_X$ is trivial), then for any invertible sheaf $\mathcal{L}$ we get

$$\chi(X, \mathcal{L}) = \frac{1}{n!} \deg c_1(\mathcal{L})^n,$$

which is the usual formulation of Riemann–Roch on abelian varieties.

We refer to Hartshorne [H], Appendix A4, to see how the Hirzebruch Riemann–Roch theorem implies the more classical Riemann–Roch theorem on curves and surfaces, except that Hartshorne's references to the Chow ring should be replaced by references to $\mathbf{Q}\mathrm{Gr}\,K$.

Theorem 7.5. *Let $f: X \to Y$ be a regular imbedding of codimension d, $\mathcal{E}$ a locally free sheaf of rank r on X. Let $e = [\mathcal{E}]$ and $q = [\mathcal{C}^\vee_{X/Y}]$ in $K(X)$. Then*

$$c(f_K(e)) = 1 + f_G(P_{r,d}(e, q)) \qquad \text{in } \mathbf{Q}\mathrm{Gr}\,K(Y).$$

Here $P_{r,d}$ is the universal polynomial defined in Chapter II, §4.

Proof. This follows from Theorem 4.3 of Chapter II and the deformation to the normal bundle, as in Lemma 6.1, 6.2, and 7.2.

Remark. Since the covariant map f_G is defined only for $G = \mathbf{Q}\mathrm{Gr}\,K$, after tensoring with $\mathbf{Q}$, the preceding theorem is not a Riemann–Roch theorem "without denominators". In other theories, when $\mathrm{Gr}\,K$ is replaced by the Chow ring, then the same type of proof does give a Riemann–Roch without denominator. For relations among K, $\mathrm{Gr}\,K$, and rational equivalence, we refer to [SGA 6], [BFM 1], or [F 2].

The Adams Riemann–Roch theorem for imbeddings in §6 required no denominators. The next theorem gives the general version for ψ^j, valid after inverting j. By precisely the same reasoning as for Grothendieck Riemann–Roch (Theorem 7.3), we have:

Theorem 7.6 (Adams Riemann–Roch). *For any $f: X \to Y$ in $\mathfrak{C}$, Riemann–Roch holds for f with respect to*

$$(K, \psi^j, \mathbf{Z}[1/j] \otimes K),$$

with multiplier

$$\theta^j(T_f^\vee)^{-1}.$$

Appendix. Non-connected Schemes

For a Noetherian scheme X which may not be connected, to specify a locally free sheaf $\mathscr{E}$ on X is the same as giving a locally free sheaf $\mathscr{E}_\alpha$ on each connected component X_α of X; each $\mathscr{E}_\alpha$ has constant rank, but these ranks may differ from component to component. When one defines $K(X)$ as in §1, one has a canonical isomorphism of rings giving a **product decomposition**

$$K(X) \cong \prod_\alpha K(X_\alpha).$$

Each $K(X_\alpha)$ is a λ-ring, but $K(X)$ is not a λ-ring as we have defined it in Chapter I. The augmentation (i.e. rank homomorphism) is a sum of the augmentation on each $K(X_\alpha)$:

$$\varepsilon: K(X) \to \mathbf{Z}^{\pi_0(X)},$$

where $\pi_0(X)$ is the set of connected components of X.

Rather than develop a theory of λ-rings with such augmentations, we have preferred to concentrate on the connected case. At any rate, any assertions for general X follow readily from the product decomposition. For example, the operations λ^i, γ^i, and ψ^i operate on $K(X)$ via their action on each $K(X_\alpha)$. For the γ-filtration,

$$F^n K(X) = \prod_\alpha F^n K(X_\alpha).$$

Hence, if $\dim X \leq d$, then $F^{d+1}K(X) = 0$, and

$$\mathrm{ch}: \mathbf{Q}K(X) \to \mathbf{Q}\mathrm{Gr}\, K(X)$$

is an isomorphism. For a morphism $f: X \to Y$, f maps each connected component X_α of X to some component $Y_{\beta(\alpha)}$ of Y. The pull-back f^K and push-forward f_K are defined by

$$f^K\left(\prod_\beta y_\beta\right) = \prod_\alpha f^K(y_{\beta(\alpha)}),$$

$$f_K\left(\prod_\alpha x_\alpha\right) = \prod_\beta\left(\prod_{\beta(\alpha)=\beta} f_K(x_\alpha)\right).$$

The Grothendieck and Adams Riemann–Roch theorems of the preceding two sections are valid without change for schemes which may not be connected; indeed, they follow immediately from the connected cases and the product decomposition.

CHAPTER VI

An Intersection Formula. Variations and Generalizations

The first point of this chapter is to develop a commutative diagram similar to that of the Riemann–Roch theorems, and called the Intersection Formula for the K-functor. In particular, this will show how the product in the ring $K(X)$ relates to the geometric intersection of subschemes of X. From this intersection formula for K we deduce a corresponding formula for Gr K, which is analogous to the "excess intersection formula" of [FM], cf. [F 2], Theorem 6.3. Special cases of the intersection formula are contained in [SGA 6] and [Man], but the general version given here for K-theory seems to be new. Our proof eliminates the use of Tor, and gives another striking illustration of the deformation formalism of Chapter II.

We then introduce the Grothendieck group of coherent sheaves on a scheme, and show how this group relates with the K-groups studied in Chapter V. In particular, this involves looking at two separate functors, $K^{\cdot}$ and $K_{\cdot}$ which are contravariant and covariant respectively. The functor $K^{\cdot}$ is the Grothendieck group of locally free sheaves as before, but $K_{\cdot}$ is the Grothendieck group of coherent sheaves. Our discussion sheds further light on the Grothendieck filtration by relating it to more geometric properties.

We shall apply special cases of the Intersection Formula (known previously) to determine the structure of K of a blow up. This is both a complement to the K-theory of blow ups, and also illustrates geometric techniques. We follow [SGA 6] and [Man], §15, with some simplifications. We thought it would be useful for the reader to see how this material follows directly from what we have already done. Note that in both [SGA 6] and [Man] the calculation of K of a blow up played an important role in the proof of Riemann–Roch theorem, while our proofs required no such calculation.

Next we discuss a filtration for $K_{\cdot}$ and relate it to the filtration for $K^{\cdot}$ when comparable. This gives more geometric insight into the Grothendieck filtration and topological filtration.

The groups $K^{\cdot}$ and $K_{\cdot}$ and their graded groups are also basic for an extension of Riemann–Roch to schemes with arbitrary singularities. We state this singular Riemann–Roch without proof. Similarly, in the rest of the chapter, we indicate other related results of a "Riemann–Roch" nature, especially in the context of schemes, where one can use some of the formalism or results of Chapters I–V. We make no attempt to survey the extensive literature in this active area, however. In particular we ignore recent Riemann–Roch theorems for analytic spaces, for arithmetic surfaces, or involving higher K-theory, as well as relations with rational equivalence and intersection theory going beyond what we did in §3. We refer to the literature for most of the proofs. The reader may also find a more general and powerful formalism in [FM].

VI §1. The Intersection Formula

Throughout this section, we work with the same objects as in the category $\mathfrak{C}$ *of Chapter* VI, §6, §7 *namely connected schemes quasi-projective over an affine Noetherian base. Not all morphisms are subject to the same restrictions, however, and the context will make the restrictions precise.*

We shall be concerned with a fibre square

FS 1.

$$\begin{array}{ccc} X_1 & \xrightarrow{f_1} & Y_1 \\ \downarrow{\scriptstyle\psi} & & \downarrow{\scriptstyle\varphi} \\ X & \xrightarrow[f]{} & Y \end{array}$$

Unless otherwise specified, the vertical morphisms ψ, φ *are morphisms of schemes, but the horizontal morphisms* f, f_1 *are assumed to be regular morphisms. We let* d, d_1 *be their respective codimensions.*

Remarks. Since a regular morphism is one which can be factored into a local complete intersection imbedding, and a projective bundle projection, it follows that a regular morphism is proper.

Even though we make no restrictive assumptions on φ, ψ we note that the contravariant maps φ^K and ψ^K are defined on the K-groups. We needed restrictions only to define the covariant maps.

If we factor f into a regular imbedding $i\colon X \to P$ followed by a projective bundle projection $p\colon P \to X$, we obtain a fibre diagram

FS 2.

$$\begin{array}{ccccc} X_1 & \xrightarrow{i_1} & P_1 & \xrightarrow{p_1} & Y_1 \\ \downarrow{\scriptstyle\psi} & & \downarrow{\scriptstyle\eta} & & \downarrow{\scriptstyle\varphi} \\ X & \xrightarrow[i]{} & P & \xrightarrow[p]{} & Y \end{array}$$

with $p_1 \circ i_1 = f_1$ and i_1 a regular imbedding. Since the ideal sheaf of X in P generates the ideal sheaf of X_1 in P_1, the left square yields a surjection

$$\psi^* \mathscr{C}_{X/P} \to \mathscr{C}_{X_1/P_1} \to 0.$$

We let $\mathscr{E}$ be the kernel, which is a locally free sheaf on X_1, so we have the exact sequence

$$(1.1) \qquad 0 \to \mathscr{E} \to \psi^* \mathscr{C}_{X/P} \to \mathscr{C}_{X_1/P_1} \to 0.$$

Arguing as in the proof of Proposition 7.1, Chapter V, one verifies easily that $\mathscr{E}$ is independent of the factorization of f. We may call $\mathscr{E}$ the **excess conormal sheaf** for the diagram **FS 1.** We let $e = [\mathscr{E}]$ be its class in $K(X_1)$, so we have

$$e = [\psi^* \mathscr{C}_{X/P}] - [\mathscr{C}_{X_1/P_1}].$$

The rank m of $\mathscr{E}$ is called the **excess dimension**

$$m = d - d_1.$$

If f, φ are regular imbeddings, then X_1 is the intersection of Y_1 and X in Y. Classically, this **intersection** is called **proper** if the excess dimension is equal to 0.

Proposition 1.1. *If the excess dimension is* 0, *that is,* f, f_1 *have the same codimension, then the following diagram commutes*:

$$\begin{array}{ccc} K(X) & \xrightarrow{f_K} & K(Y) \\ \downarrow{\scriptstyle \psi^K} & & \downarrow{\scriptstyle \varphi^K} \\ K(X_1) & \xrightarrow[f_{1K}]{} & K(Y_1) \end{array}$$

Proof. Factoring f as above, it suffices to prove the proposition when f is a closed imbedding or a projective bundle projection. The imbedding case was proved in Chapter V, Proposition 4.5. For the projection case, suppose $X = \mathbf{P}(\mathscr{G})$ with $\mathscr{G}$ locally free on Y, and $f\colon \mathbf{P}(\mathscr{G}) \to Y$ is the projection. Then

$$X_1 = \mathbf{P}(\varphi^* \mathscr{G}) \qquad \text{and} \qquad f_1\colon X_1 \to Y_1 \quad \text{is the projection.}$$

To prove the assertion, it will suffice to prove:

Lemma 1.2. *Let $X = \mathbf{P}(\mathscr{G})$ and let $f: X \to Y$ be the projection. Let $\mathscr{F}$ be a regular locally free sheaf on X. Then $\psi^*\mathscr{F}$ is regular on X_1, and*

$$\varphi^* f_* \mathscr{F} \approx f_{1*}\psi^*\mathscr{F}.$$

Proof. By Chapter V, (2.5) there is a canonical resolution of $\mathscr{F}$:

$$0 \to (f^*\mathscr{T}_r)(-r) \to \cdots \to (f^*\mathscr{T}_2)(-1) \to f^*\mathscr{T}_0 \to \mathscr{F} \to 0.$$

with locally free sheaves $\mathscr{T}_i = \mathscr{T}_i(\mathscr{F})$ on Y, $\mathscr{T}_0 = f_*\mathscr{F}$. Since these sheaves are locally free, the pull-back of this sequence by ψ^* is exact on X_1. Since

$$\psi^* f^* = f_1^* \varphi^* \qquad \text{and} \qquad \psi^*\mathscr{O}_{\mathbf{P}\mathscr{G}}(1) = \mathscr{O}_{\mathbf{P}(\varphi^*\mathscr{G})}(1),$$

we get an exact sequence on X_1:

$$0 \longrightarrow (f_1^*\varphi^*\mathscr{T}_r)(-r) \xrightarrow{d_r} \cdots \xrightarrow{d_1} f_1^*\varphi^*\mathscr{T}_0 \xrightarrow{d_0} \psi^*\mathscr{F} \longrightarrow 0.$$

The lemma follows readily from this resolution, namely let

$$\mathscr{Z}_i = \operatorname{Ker} d_i,$$

so there are short exact sequences for $i > 0$:

$$\text{(A)} \qquad 0 \to \mathscr{Z}_i \to (f_1^*\varphi^*\mathscr{T}_i)(-i) \to \mathscr{Z}_{i-1} \to 0$$

and for $i = 0$,

$$\text{(B)} \qquad 0 \to \mathscr{Z}_0 \to f_1^*\varphi^*\mathscr{T}_0 \to \psi^*\mathscr{F} \to 0.$$

Starting with $\mathscr{Z}_r = 0$ one uses the long exact cohomology sequence of (A) to show by descending induction that $\mathscr{Z}_i(1)$ is regular, and $f_{1*}\mathscr{Z}_i = 0$ for all $i \geqq 0$. From (B) one deduces that $\psi^*\mathscr{F}$ is regular, and that

$$f_{1*}(f_1^*\varphi\mathscr{T}_0) \to f_{1*}\psi^*\mathscr{F}$$

is an isomorphism. Since $\mathscr{T}_0 = f_*\mathscr{F}$ it follows that

$$\varphi^* f_*\mathscr{F} = f_{1*}f_1^*(\varphi^* f_*\mathscr{F}) = f_{1*}(f_1^*\varphi^*\mathscr{T}_0),$$

which proves that $\varphi^* f_*\mathscr{F} \approx f_{1*}\psi^*\mathscr{F}$. This proves the lemma.

Theorem 1.3 (Intersection Formula). *Given a fiber square* **FS 1** *with excess conormal sheaf* $\mathscr{E}$, *let* e *be the class of* $\mathscr{E}$ *in* $K(X_1)$. *Then the following diagram is commutative*:

$$\begin{array}{ccc} K(X) & \xrightarrow{f_K} & K(Y) \\ \downarrow{\scriptstyle \lambda_{-1}(e)\psi^K} & & \downarrow{\scriptstyle \varphi^K} \\ K(X_1) & \xrightarrow[f_{1K}]{} & K(Y_1) \end{array}$$

Before presenting the proof, we record some special cases.

1.3.1. Excess Dimension 0 (Proper Intersection). In this case, $\mathscr{E} = 0$, $\lambda_{-1}(e) = 1$, and the formula reduces to that of Proposition 1.1.

1.3.2. Self Intersection Formula. This is the other extreme, when

$$Y_1 = X, \ \varphi = f,$$

and f is a regular imbedding. Then $X_1 = X$, $E = \mathscr{C}_{X/Y}$ is the conormal sheaf, and the formula reads

$$f^K f_K(x) = \lambda_{-1}(c)x,$$

where $c = [\mathscr{C}_{X/Y}]$.

1.3.3. Blow Up or Key Formula. In this case, f is a regular imbedding, and $\varphi\colon Y_1 \to Y$ is the blow up of a regular imbedding $f\colon X \to Y$, so $Y_1 = \mathrm{Bl}_X(Y)$. Then

$$X_1 = \mathbf{P}(\mathscr{C}_{X/Y}), \qquad \mathscr{C}_{X_1/Y_1} \approx \mathscr{O}_{\mathbf{P}\mathscr{C}_{X/Y}}(1),$$

and the exact sequence (1.1) is the universal exact sequence

$$0 \to \mathscr{E} \to \psi^*\mathscr{C}_{X/Y} \to \mathscr{O}_{\mathbf{P}\mathscr{C}_{X/Y}}(1) \to 0.$$

We usually let $\ell = [\mathscr{O}_{\mathbf{P}\mathscr{C}_{X/Y}}(1)] = [\mathscr{C}_{X_1/Y_1}]$, and then

$$e = \psi^K(c) - \ell,$$

where $c = [\mathscr{C}_{X/Y}]$.

We may now pass to the graded case.

Corollary 1.4. *Let G be the functor $G = \mathbf{Q}\mathrm{Gr}\, K$. With assumptions as in Theorem 1.3, we have a commutative diagram*

$$\begin{array}{ccc} G(X) & \xrightarrow{f_G} & G(Y) \\ {\scriptstyle c_m(e^\vee)\psi^G}\downarrow & & \downarrow{\scriptstyle \varphi^G} \\ G(X_1) & \xrightarrow[f_{1G}]{} & G(Y_1) \end{array}$$

Proof. Given $x \in G^k X$ choose a representative $\tilde{x}$ for x in $\mathbf{Q}F^k KX$. By Proposition 2.1(i) of Chapter III, $\lambda_{-1}(e)$ is a representative for $c_m(e^\vee)$ in $F^m KX_1$. By Theorem 1.3,

$$f_{1K}(\lambda_{-1}(e)\psi^K \tilde{x}) = \varphi^K f_K(\tilde{x})$$

in $\mathbf{Q}F^{k+d}K(Y_1)$, and this represents the required equation in $G^{k+d}Y_1$, thus proving the corollary.

As in the non-graded result, we have the three special cases:

1.4.1. Excess Dimension 0. The formula reads

$$f_{1G}\psi^G = \varphi^G f_G.$$

1.4.2. Self Intersection Formula. If $\varphi = f$ is a regular imbedding, then

$$f^G f_G(x) = c_m(e^\vee)x,$$

with $m = d$. In this case, $\mathscr{E}^\vee$ is the normal sheaf.

1.4.3. Blow Up or Key Formula. Here f is a regular imbedding,

$$Y_1 = \mathrm{Bl}_X(Y),$$

and $\mathscr{E}$ is the universal subsheaf on $X_1 = \mathbf{P}(\mathscr{C}_{X/Y})$. Then

$$f_{1G}(c_{d-1}(e^\vee)\psi^G(x)) = \varphi^G f_G(x).$$

Remark. The proof we shall give for the theorem can be modified slightly to prove the corollary directly: one replaces K by G, $\lambda_{-1}(e)$ by $c_m(e^\vee)$, and Proposition 4.4(a) by Proposition 4.4(b). This proof has the advantage that it works in other contexts, such as rational equivalence theory or other cohomology theories where it is not necessary to tensor with $\mathbf{Q}$.

VI §2. Proof of the Intersection Formula

Factoring f into a regular imbedding followed by a projection, it suffices as usual to prove the formula in each case. The projection case is covered by Proposition 1.1, so we may assume that f is an imbedding.

As in Chapter II, we meet a situation which splits in two parts, one formal the other not. This involves deforming an imbedding to an "elementary imbedding" (suitably defined for the present application), proving the formula formally for "elementary imbeddings", and showing that if the formula is true for a morphism, then it is true for a "deformation", suitably axiomatized.

So we start with the axiomatization. Let $\mathfrak{C}$ be a category. We have already observed the need for two kinds of morphisms, so we have to build this into the axioms. Hence we suppose given for each two objects a subset of their morphisms, called **restricted morphisms**, such that the restricted morphisms form a subcategory.

By a **λ-ring functor** K we now mean that the association $X \mapsto K(X)$ is contravariant for all morphisms, also covariant for restricted morphisms, and satisfies the projection formula for restricted morphisms. Let K be such a functor.

In Chapter II, §3 we defined an elementary imbedding with respect to K. Given a morphism $f: X \to Y$ the surjectivity of $f^K: K(Y) \to K(X)$ will here come from the fact that f is a section of a morphism $p: Y \to X$. The other condition was that $f_K(1) = \lambda_{-1}(q)$ for some element $q \in K(Y)$. Both these conditions are going to play a role.

In addition, let

$$\begin{array}{ccc} X_1 & \xrightarrow{f_1} & Y_1 \\ {\scriptstyle\psi}\downarrow & & \downarrow{\scriptstyle\varphi} \\ X & \xrightarrow[f]{} & Y \end{array}$$

be a commutative square in $\mathfrak{C}$ with f, f_1 restricted. We shall say that the **intersection formula holds** for this square with **multiplier** $\lambda_{-1}(e)$ for some element $e \in K(X_1)$ if the following diagram commutes:

$$\begin{array}{ccc} K(X) & \xrightarrow{f_K} & K(Y) \\ {\scriptstyle\lambda_{-1}(e)\psi^K}\downarrow & & \downarrow{\scriptstyle\varphi^K} \\ K(X_1) & \xrightarrow[f_{1K}]{} & K(Y_1) \end{array}$$

We shall say that a commutative square in $\mathfrak{C}$ as above is **elementary** if the following properties are satisfied:

ES 1. The morphisms f, f_1 are sections of morphisms

$$p: Y \to X \qquad \text{and} \qquad p_1: Y_1 \to X_1$$

such that $p \circ \varphi = \psi \circ p_1$.

ES 2. There exists elements $q \in K(Y)$ and $q_1 \in K(Y_1)$ such that

$$f_K(1) = \lambda_{-1}(q) \qquad \text{and} \qquad f_{1K}(1) = \lambda_{-1}(q_1).$$

ES 3. There exists an element $e \in K(X_1)$ such that

$$\varphi^K \lambda_{-1}(q) = p_1^K \lambda_{-1}(e) \cdot \lambda_{-1}(q_1).$$

Proposition 2.1. *Assume that the commutative square is elementary. Then the intersection formula holds with multiplier $\lambda_{-1}(e)$.*

Proof. For $x \in K(X)$ we have:

	reasons
$\varphi^K f_K(x) = \varphi^K f_K(f^K p^K x)$	**ES 1**, $p \circ f = \mathrm{id}$
$= \varphi^K(p^K x \cdot f_K(1))$	projection formula
$= \varphi^K(p^K x \cdot \lambda_{-1}(q))$	**ES 2**
$= p_1^K(\psi^K x) \cdot \varphi^K \lambda_{-1}(q)$	**ES 1**, $p \circ \varphi = \Psi \circ p_1$
$= p_1^K(\psi^K x) \cdot p_1^K \lambda_{-1}(e) \cdot \lambda_{-1}(q_1)$	**ES 3**
$= p_1^K(\psi^K x) \cdot p_1^K \lambda_{-1}(e) \cdot f_{1K}(1)$	**ES 2**
$= f_{1K}(f_1^K p_1^K(\psi^K x) \cdot \lambda_{-1}(e))$	projection formula
$= f_{1K}(\psi^K x \cdot \lambda_{-1}(e))$	**ES 1**, $p_1 \circ f_1 = \mathrm{id}$

This proves the proposition.

Geometric Construction of an Elementary Square

We shall now construct a situation in the geometric category which satisfies the axioms of an elementary square.

We suppose given:

a morphism $\psi : X_1 \to X$;

locally free sheaves $\mathscr{F}$ on X and $\mathscr{F}_1$ on X_1 and a surjection

$$\alpha: \psi^* \mathscr{F} \to \mathscr{F}_1.$$

With such data, we let

$$Y = \mathbf{P}(\mathscr{F} \oplus \mathcal{O}_X) \qquad \text{and} \qquad Y_1 = \mathbf{P}(\mathscr{F}_1 \oplus \mathcal{O}_{X_1})$$

together with their projections

$$p: Y \to X \qquad \text{and} \qquad p_1: Y_1 \to X_1.$$

Finally, we let

$$f: X \to Y \qquad \text{and} \qquad f_1: X_1 \to Y_1$$

be the zero section imbeddings.

The homomorphism α induces a morphism

$$\varphi: \mathbf{P}(\mathscr{F}_1 \oplus \mathcal{O}_{X_1}) \to \mathbf{P}(\mathscr{F} \oplus \mathcal{O}_X)$$

giving a *fibre square* as in **FS 1**. In this case, $\mathscr{F}$ and $\mathscr{F}_1$ are the conormal sheaves to f and f_1 respectively, so the excess sheaf $\mathscr{E}$ is the kernel of α. We let $e = [\mathscr{E}]$.

Moreover, if $\mathscr{Q}$ and $\mathscr{Q}_1$ are the universal hyperplane sheaves on Y and Y_1 as in Chapter IV, §1, and q, q_1 are their respective classes in $K(Y)$ and $K(Y_1)$, then **ES 1** is trivially satisfied, and Proposition 4.3(a) of Chapter V shows that **ES 2** is satisfied.

We shall now prove **ES 3**. We have a commutative diagram with exact rows and columns on Y_1:

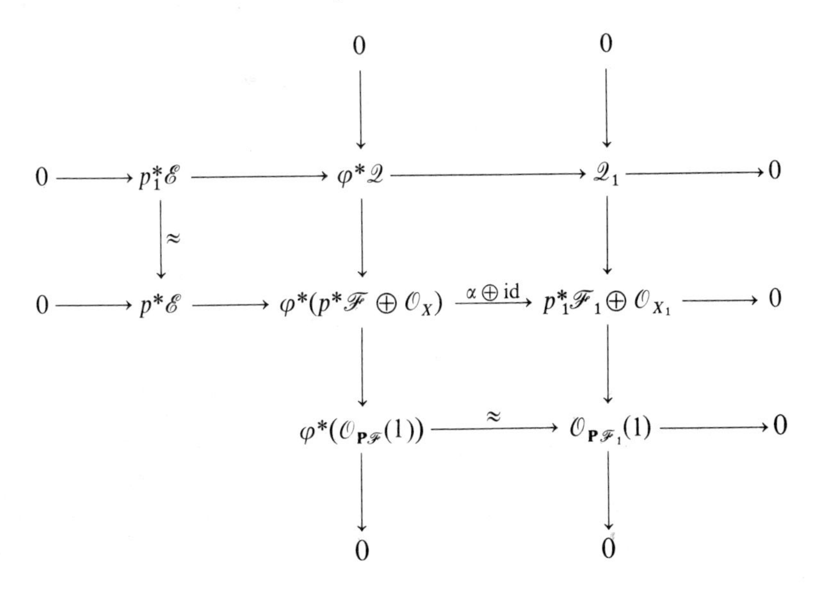

From the top row, we deduce **ES 3** as desired.

We return to the proof of Theorem 1.3 when f is a regular imbedding. The idea is to use a deformation diagram as in Chapter II, §1. Again, since there is only one ring K here, we restate all hypothesis *ab ovo*, and we first describe the axiomatization. In part this amounts to repeating the **BD** conditions in the special case when $A = K$ and $\rho = \text{id}$. The additional feature amounts to saying that the deformation diagram is functorial. On the other hand, we do not need all the **BD** axioms except for certain maps, so we list just the properties that we need.

So again we let $\mathfrak{C}$ be an arbitrary category with restricted morphisms, and we let K be a λ-ring functor on $\mathfrak{C}$. Suppose given a commutative square in $\mathfrak{C}$ with f, f_1 restricted:

$$\begin{array}{ccc} X_1 & \xrightarrow{f_1} & Y_1 \\ {\scriptstyle\psi}\downarrow & & \downarrow{\scriptstyle\varphi} \\ X & \xrightarrow[f]{} & Y \end{array}$$

We shall say that this square admits a **basic deformation** to a square

$$\begin{array}{ccc} X_1 & \xrightarrow{f'_1} & Y'_1 \\ {\scriptstyle\psi}\downarrow & & \downarrow{\scriptstyle\varphi'} \\ X & \xrightarrow[f']{} & Y' \end{array}$$

if there exist morphisms as shown on the following diagram called the **deformation cube**:

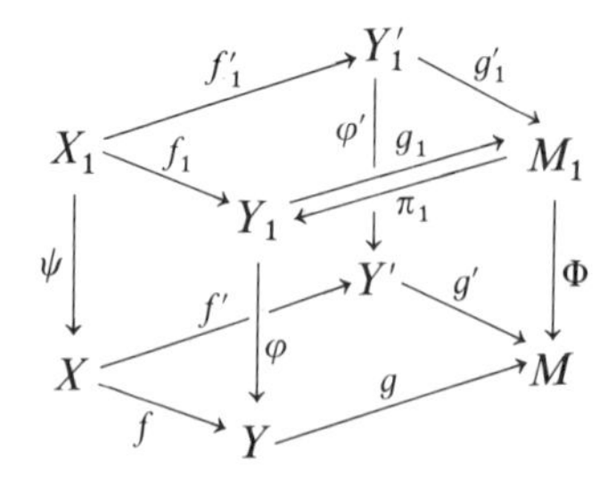

such that all the horizontal morphisms are restricted, and there exists a finite number of restricted morphisms

$$h_{1\nu}: C_\nu \to M_1$$

with integers $m_\nu \in \mathbf{Z}$, satisfying the following conditions:

SBD 1. For each $x \in K(X)$ there exists $z \in K(M)$ such that

$$f_K(x) = g^K(z) \quad \text{and} \quad f'_K(x) = g'^K(z).$$

SBD 2. $g_{1K}(1) = g'_{1K}(1) + \sum m_\nu h_{1\nu K}(1).$

SBD 3. For each $z \in K(M)$ as in **SBD 1**, and all ν, we have

$$h^K_{1\nu}\Phi^K(z) = 0.$$

SBD 4. The four vertical faces going around the cube are commutative;

g_1 is a section of π_1 and $\pi_1 \circ g'_1 \circ f'_1 = f_1$.

Proposition 2.2. *Suppose given a commutative square with restricted horizontal morphisms, and that this square admits a basic deformation as above. If the intersection formula holds for the square of a basic deformation*

$$\begin{array}{ccc} X_1 & \xrightarrow{f'_1} & Y'_1 \\ \downarrow{\scriptstyle\psi} & & \downarrow{\scriptstyle\varphi'} \\ X & \xrightarrow[f']{} & Y' \end{array}$$

then the intersection formula holds for the given square with the same multiplier.

Proof. The proof consists in following the same pattern as the analogous statement of the Riemann–Roch formula, Theorem 1.3 of Chapter II. We just go around the cube as follows. Given $x \in K(X)$ choose $z \in K(M)$ as in **SBD 1**. Then:

	reasons
$g_{1K}(\varphi^K f_K(x)) = g_{1K}(\varphi^K g^K z)$	**SBD 1**
$= g_{1K}(g_1^K \Phi^K z)$	**SBD 4**
$= g_{1K}(1)\Phi^K z$	projection formula
$= g'_{1K}(1)\Phi^K z + \sum m_\nu h_{1\nu K}(1)\Phi^K z$	**SBD 2**
$= g'_{1K}(g'^K_1\Phi^K z) + \sum m_\nu h_{1\nu K}(h^K_{1\nu}(\Phi^K z))$	projection formula
$= g'_{1K}\varphi'^K g'^K z + 0$	**SBD 4** and **SBD 3**
$= g'_{1K}\varphi'^K f'_K(x)$	**SBD 1**
$= g'_{1K} f'_{1K}(\lambda_{-1}(e)\psi^K(x))$	intersection formula

We now apply π_{1K}. Since g_1 is a section of π_1, we have

$$\pi_{1K}g_{1K} = \text{id},$$

so we find:

$$\begin{aligned}\varphi^K f_K(x) &= \pi_{1K}g'_{1K}f'_{1K}(\lambda_{-1}(e)\psi^K(x))\\ &= f_{1K}(\lambda_{-1}(e)\psi^K(x)) \qquad \text{by } \mathbf{SBD\ 4}.\end{aligned}$$

This concludes the proof.

All that remains to be done to finish the proof of Theorem 1.3 (the Intersection Formula in the geometric context) is to prove:

Proposition 2.3. *Given a fibre square* **FS 1**, *with* f, f_1 *assumed to be regular imbeddings. There exists a basic deformation of this square to an elementary square.*

Proof. We already know that regular imbeddings $f: X \to Y$ and $f_1 : X_1 \to Y_1$ can be deformed to their normal bundles. We now note that the construction of this deformation is *functorial.* In Chapter IV, §5 we constructed from f a diagram

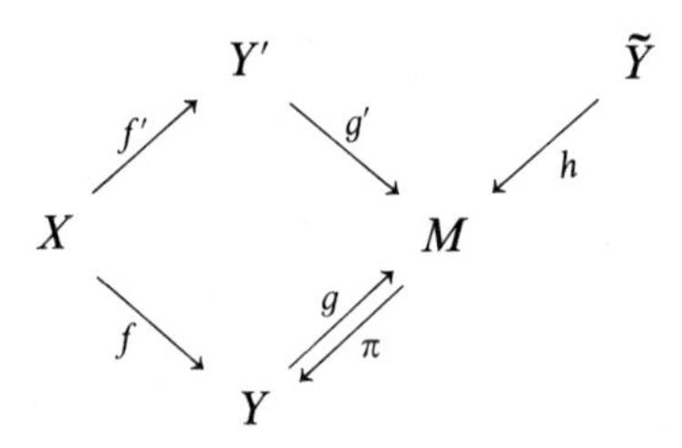

Given the morphism $\varphi: Y_1 \to Y$ and fibre square **FS 1**, we obtain a similar square for $f_1: X_1 \to Y_1$ and induced vertical morphisms giving rise to the deformation cube:

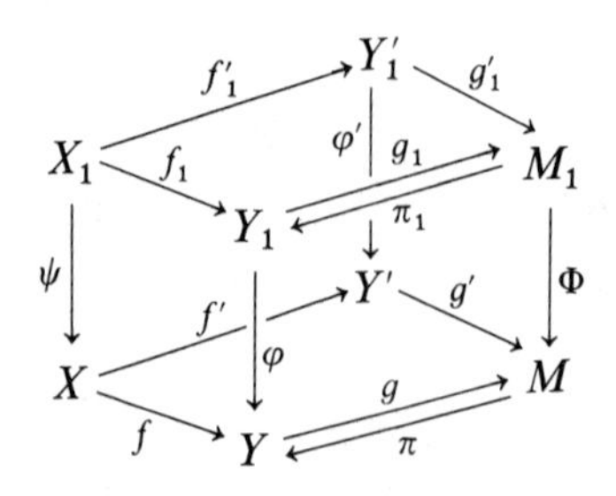

The morphism

$$\Phi : M_1 \to M$$

is the morphism of blow ups

$$\mathrm{Bl}_{X_1 \times \infty}(Y_1 \times \mathbf{P}^1) \to \mathrm{Bl}_{X \times \infty}(Y \times \mathbf{P}^1)$$

induced by

$$\varphi \times \mathrm{id} : Y_1 \times \mathbf{P}^1 \to Y \times \mathbf{P}^1.$$

Conditions **SBD 4** (the commutativity properties) are then automatically satisfied, and of course we have some others not listed in **SBD**, like

$$\pi \circ g = \mathrm{id}_Y \qquad \text{and} \qquad \pi \circ g' \circ f' = f,$$

which had not been necessary in the proof of Proposition 2.2.

The left back vertical square is then an "elementary square" satisfying **ES 1**, **ES 2**, **ES 3**, as constructed previously:

$$\begin{array}{ccl} X_1 & \xrightarrow{f'_1} & Y'_1 = \mathbf{P}(\mathscr{C}_{X_1/Y_1} \oplus \mathscr{O}_{X_1}) \\ \downarrow & & \downarrow \\ X & \xrightarrow[f']{} & Y' = \mathbf{P}(\mathscr{C}_{X/Y} \oplus \mathscr{O}_X) \end{array}$$

and the Intersection Formula holds by Proposition 2.1.

From the deformation to the normal bundle of Chapter IV, §5 we also have the residual schemes Y and Y_1 with their imbeddings in M_1 and M respectively, and an induced morphism between them as shown on the following commutative diagram:

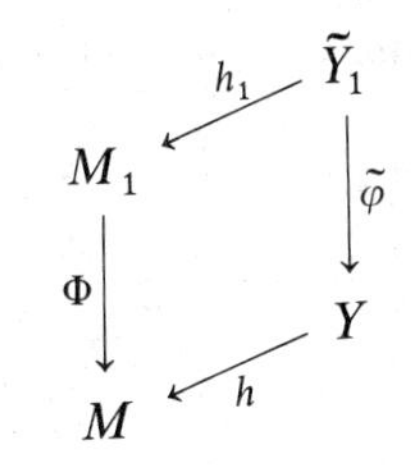

We shall also need the imbedding

$$h' : Y' \cap \tilde{Y} \to M$$

and the corresponding commutative square

$$\begin{array}{ccc} Y_1' \cap \tilde{Y}_1 & \xrightarrow{h_1'} & M_1 \\ {\scriptstyle \tilde{\varphi}'}\downarrow & & \downarrow{\scriptstyle \Phi} \\ Y' \cap \tilde{Y} & \xrightarrow[h']{} & M \end{array}$$

The top deformation together with Proposition 4.4(a) of Chapter V gives the equation

$$g_{1K}(1) = g'_{1K}(1) + h_{1K}(1) - h'_{1K}(1).$$

The construction of the basic deformation for f being the same as in the part of this book dealing with the Riemann–Roch theorem, we know from Lemma 6.1 of Chapter V that given $x \in K(X)$, there exists $z \in K(M)$ such that

$$f_K(x) = g^K(z), \qquad f'_K(x) = g'^K(z),$$

$$h^K(z) = 0, \qquad h'^K(z) = 0.$$

Then

$$h_1^K \Phi^K(z) = \tilde{\varphi}^K h^K(z) = 0,$$

and similarly

$$h_1'^K \Phi^K(z) = \varphi'^K h'^K(z) = 0.$$

This proves **SBD 2** and **SBD 3**, and concludes the proof of all the **SBD** conditions. It also concludes the proof of Proposition 2.3 and of Theorem 1.3.

VI §3. Upper and Lower K

In this section, $\mathfrak{C}$ denotes a category of Noetherian schemes, each of which has an ample invertible sheaf. For example, $\mathfrak{C}$ may be the category of quasi-projective schemes over a fixed affine Noetherian base scheme. Morphisms are arbitrary scheme morphisms.

The purpose of this section is to introduce two different K-functors. Among other things, these two functors make it possible to deal with more general singularities than have been considered up to now. For X in $\mathfrak{C}$ we let

$K^{\cdot}(X) =$ Grothendieck group of locally free sheaves on X;

$K_{\cdot}(X) =$ Grothendieck group of coherent sheaves on X.

We denoted $K^{\cdot}(X)$ by $K(X)$ before, but now shall view $K^{\cdot}(X)$ as a contravariant functor with respect to *all* scheme morphisms. We let

$$\delta : K^{\cdot}X \to K_{\cdot}X$$

be the homomorphism induced by the inclusion of $\mathfrak{V}_X$ in the category of coherent sheaves on X. This homomorphism is called the **Poincaré-homomorphism**.

Proposition 3.1. *If X is regular, then δ is an isomorphism.*

Proof. Over a regular local ring, every finitely generated module has finite homological dimension ([Mat], 18C, Theorem 45, Serre's Theorem); so if X is regular then every coherent sheaf on X has a finite locally free resolution using the basic condition $(*)$ of Chapter V, §4 and the introductory remarks of that chapter. Hence δ is an isomorphism by Proposition 4.1 of Chapter V.

For a regular scheme X, we may use δ to identify $K^{\cdot}X$ with $K_{\cdot}X$, and we write

$$K(X) = K^{\cdot}X = K_{\cdot}X.$$

Next we consider a useful exact sequence which shows the advantage of dealing with $K_{\cdot}$ in certain contexts. Let

$$i : X \to Y$$

be a closed imbedding. If $\mathscr{F}$ is a coherent sheaf on X, then $i_*\mathscr{F}$ is the sheaf on Y obtained by extending $\mathscr{F}$ to 0 outside X. Then i_* is an exact functor, and therefore induces a homomorphism

$$i_{K_{\cdot}} : K_{\cdot}X \to K_{\cdot}Y$$

by

$$i_{K_{\cdot}}[\mathscr{F}] = [i_*\mathscr{F}].$$

On the other hand, let $j : U \to Y$ be the inclusion of an open subscheme U of a scheme Y. There is a restriction homomorphism

$$j^{K_{\cdot}} : K_{\cdot}(Y) \to K_{\cdot}(U),$$

which takes the class $[\mathscr{F}]$ of a coherent sheaf $\mathscr{F}$ on Y to the class $[\mathscr{F} | U]$ of the restriction of $\mathscr{F}$ to U.

Proposition 3.2. *Let $i: X \to Y$ be the inclusion of a closed subscheme, let U be the complement of X in Y, and let $j: U \to Y$ be the inclusion. Then the sequence*

$$K.(X) \xrightarrow{i_{K.}} K.(Y) \xrightarrow{j^{K.}} K.(U) \longrightarrow 0$$

is exact.

Proof. It is obvious from the definitions that the composite is zero, so there is a homomorphism

$$K.(Y)/\mathrm{Im}(i_{K.}) \to K.(U).$$

To prove that his homomorphism is an isomorphism, we use Appendix 3.5 and 3.6 that a coherent sheaf $\mathscr{F}$ on U is the restriction of some coherent sheaf $\tilde{\mathscr{F}}$ on Y, and any short exact sequence of coherent sheaves on U is the restriction of an exact sequence on Y. Assigning $[\tilde{\mathscr{F}}]$ to $[\mathscr{F}]$ then determines a homomorphism

$$K.(U) \to K.(Y)/\mathrm{Im}(i_{K.})$$

which is inverse to the above homomorphism; all we have to prove is that $[\tilde{\mathscr{F}}] \bmod \mathrm{Im}(i_{K.})$ is well defined. By Lemma 3.7 it suffices to prove that if $\mathscr{F}_1$, $\mathscr{F}_2$ are two extensions of $\mathscr{F}$ to Y with a homomorphism $\mathscr{F}_1 \to \mathscr{F}_2$ on Y which is an isomorphism on U, then

$$[\mathscr{F}_1] \equiv [\mathscr{F}_2] \mod \mathrm{Im}_{K.}.$$

But the kernel and cokernel of $\mathscr{F}_1 \to \mathscr{F}_2$ have support in the complement of U, thus proving the assertion and concluding the proof of Proposition 3.2.

The definition of $i_{K.}$ for a closed imbedding i was given *ad hoc*. We now study the covariant functoriality of $K.$ more systematically.

Let $f: X \to Y$ be a proper morphism. We define the **push-forward**

$$f_{K.}: K.(X) \to K.(Y)$$

by the formula

$$f_{K.}[\mathscr{F}] = \sum_{i \geqq 0} (-1)^i [R^i f_* \mathscr{F}].$$

The long exact cohomology sequence shows that $f_{K.}$ is well defined on $K.(X)$, and the spectral sequence for a composite shows that

$$(f \circ g)_{K.} = f_{K.} \circ g_{K.},$$

so $K_.$ is covariant for proper morphisms. For a proof, see for instance [L], Chapter IV, Theorem 9.8.

We note that if $f: X \to Y$ is a closed imbedding, then

$$R_i f_* \mathscr{F} = 0 \quad \text{for all} \quad i > 0,$$

and consequently the above definition coincides with the *ad hoc* definition given for closed imbeddings in the preceding section. Indeed, f_* is an exact functor (extension by 0 outside X), and hence $R^i f_* = 0$ for $i \geqq 1$.

The definition of $f_{K.}$ above is compatible with the previous definition of f_K whenever it is possible to compare them. More precisely:

Proposition 3.3. *The following diagram is commutative*:

$$\begin{array}{ccc} K^{\cdot}(X) & \xrightarrow{\delta} & K_.(X) \\ \downarrow{\scriptstyle f_{K^{\cdot}}} & & \downarrow{\scriptstyle f_{K.}} \\ K^{\cdot}(Y) & \xrightarrow{\delta} & K_.(Y) \end{array}$$

It suffices to prove this when f is a closed imbedding or when f is a projective bundle projection. We have just made the relevant remark for a closed imbedding. For a projection, $f_{K.}$ was defined on regular sheaves $\mathscr{F}$ by

$$f_K[\mathscr{F}] = [f_* \mathscr{F}],$$

and here again this is the same as the new definition because for regular sheaves, $R^i f_* \mathscr{F} = 0$ for $i > 0$.

Tensor product makes $K_.X$ a module over $K^{\cdot}X$

$$K^{\cdot}X \otimes K_.X \to K_.X \qquad \text{by} \qquad [\mathscr{E}] \cdot [\mathscr{F}] = [\mathscr{E} \otimes \mathscr{F}].$$

For example, the Poincaré homomorphism

$$\delta: K^{\cdot}X \to K_.X$$

takes an element x in $K^{\cdot}X$ to the element $x \cdot [\mathscr{O}_X]$ in $K_.X$.

From **R3** of Chapter V, §2 one deduces the **Projection Formula:**

Proposition 3.4. *For $f: X \to Y$ proper, $x \in K_.X$, $y \in K^{\cdot}Y$, we have*

$$f_{K.}(f^{K^{\cdot}}(y) \cdot x) = y \cdot f_{K.}(x).$$

We shall meet still another projection formula in Proposition 6.2.

Appendix. Basic Lemmas

Throughout this appendix, we let U be an open subscheme of a Noetherian scheme X.

Lemma 3.5. *Let $\mathscr{G}_U$ be a coherent sheaf on U. Then there exists a coherent sheaf $\mathscr{G}$ on X such that*

$$\mathscr{G} \mid U = \mathscr{G}_U.$$

If $\mathscr{F}$ is a quasi-coherent sheaf on X and $\mathscr{G}_U$ is given as a subsheaf of $\mathscr{F} \mid U$, then $\mathscr{G}$ can be taken as a subsheaf of $\mathscr{F}$.

Proof. We give the proof in the case of the given $\mathscr{F}$ and $\mathscr{G}_U$ subsheaf of $\mathscr{F}$. The proof in the absolute case without $\mathscr{F}$ is obtained by deleting all references to $\mathscr{F}$.

Consider all pairs $(\mathscr{G}, W)$ consisting of an open subscheme W of X and a coherent subsheaf $\mathscr{G}$ of $\mathscr{F} \mid W$ extending $(\mathscr{G}_U, U)$. Such pairs are partially ordered by inclusion of W's, and are in fact inductively ordered because the notion of a coherent sheaf is local, so the usual union over a totally ordered subfamily gives a pair dominating every element of the family. By Zorn's lemma, there exists a maximal element of the family, say $(\mathscr{G}, W)$. We reduce the proposition to the affine case as follows. If $W \neq X$ then there is an affine open subscheme $V = \mathrm{Spec}(A)$ in X such that $V \not\subset W$. Then $W \cap V$ is an open subscheme of V, and if we have the proposition in the affine case, then we extend $\mathscr{G}$ from $W \cap V$ to V, thus extending $\mathscr{G}$ to a larger subscheme than W, contradicting the maximality.

We now prove the lemma when X is affine. In that case, note that the coherent subsheaves of $\mathscr{G}_U$ satisfy the ascending chain condition. We let $\mathscr{G}_1$ be a maximal coherent subsheaf of $\mathscr{G}_U$ which admits a coherent extension $\mathscr{G}$ which is a subsheaf of $\mathscr{F}$. We want to prove that $\mathscr{G}_1 = \mathscr{G}_U$. If $\mathscr{G}_1 \neq \mathscr{G}_U$ then there exists an affine open $X_f \subset U$ and a section $s \in \mathscr{G}(X_f)$ such that $s \notin \mathscr{G}_1(X_f)$. By [H], II, Lemma 5.3, there exists n such that $f^n s$ extends to a section $s' \in \mathscr{F}(X)$, and the restriction of s' to U is in $\mathscr{F}(U)$. By the same reference, there exists a still higher power f^m such that

$$f^m(s' \mid U) = 0 \text{ in } (\mathscr{F}/\mathscr{G})(U).$$

Then $\mathscr{G}_1 + f^m s' \mathscr{O}_X$ is a coherent subsheaf of $\mathscr{F}$ which is bigger than $\mathscr{G}_1$, contradiction. This concludes the proof of Lemma 3.5.

Lemma 3.6. *A short exact sequence of coherent sheaves on U is the restriction of an exact sequence of coherent sheaves on X.*

Proof. Let

$$0 \to \mathcal{G}'_U \to \mathcal{G}_U \to \mathcal{G}''_U \to 0$$

be an exact sequence of coherent sheaves on U. By Lemma 3.5 there is a coherent extension $\mathcal{G}$ of $\mathcal{G}_U$ to X, and there is an extension $\mathcal{G}'$ to $\mathcal{G}'_U$ to a coherent subsheaf of $\mathcal{G}$ on X. We let $\mathcal{G}'' = \mathcal{G}/\mathcal{G}'$ to conclude the proof.

Lemma 3.7. *Let $\mathcal{F}$ be coherent on U and let $\mathcal{F}_1$, $\mathcal{F}_2$ be coherent on X such that their restrictions to U are isomorphic to $\mathcal{F}$. Then there exists a coherent sheaf $\mathcal{G}$ on X and homomorphisms*

$$\mathcal{G} \to \mathcal{F}_1, \qquad \mathcal{G} \to \mathcal{F}_2$$

which are isomorphism on U.

Proof. Let $\mathcal{G}_U$ be the graph of an isomorphism on U between $\mathcal{F}_1 \mid U$ and $\mathcal{F}_2 \mid U$. By Lemma 3.6 there exists a coherent subsheaf $\mathcal{G}$ of $\mathcal{F}_1 \oplus \mathcal{F}_2$ whose restriction to U is $\mathcal{G}_U$. This subsheaf $\mathcal{G}$ has the required property, the homomorphisms to $\mathcal{F}_1$ and $\mathcal{F}_2$ being the projections.

VI §4. *K* of a Blow Up

In the first part of this section, we let $f: X \to Y$ be a regular imbedding in the category $\mathfrak{C}$ of Chapter V, §6, §7. *We let*

$$\begin{array}{ccc} X_1 & \xrightarrow{f_1} & Y_1 \\ \psi \downarrow & & \downarrow \varphi \\ X & \xrightarrow[f]{} & Y \end{array}$$

be the blow up diagram of X in Y. We let $K = K^{\cdot}$ unless otherwise specified. Note that φ, ψ are regular morphisms.

The next proposition gives one more geometric result about blow ups.

Proposition 4.1. *In the blow up diagram, the map*

$$\varphi_K \varphi^K : K(Y) \to K(Y)$$

is the identity map, so $\varphi_K(1) = 1$.

Proof. The special case $\varphi_K(1) = 1$ is equivalent to the general formula by the projection formula. So we prove the special case. In fact, we shall prove it only under the assumption that X and Y are regular (so X_1, Y_1 are also regular). The general case requires the Remark following Proposition 5.1 of Chapter V, see [SGA 6], VII, Proposition 3.6. Under the regularity assumption, we have $K = K.$ and we can use the definition

$$\varphi_{K.}(1) = \sum_{i=0}^{\infty} (-1)^i [R^i \varphi_* \mathcal{O}_{Y_1}].$$

Thus it suffices to prove:

$$R^i \varphi_* \mathcal{O}_{Y_1} = \begin{cases} \mathcal{O}_Y & \text{if } i = 0, \\ 0 & \text{if } i > 0. \end{cases}$$

Let $\mathcal{I}_1$ be the ideal sheaf of X_1 in $\mathcal{O}_{Y_1}$. By Lemma 4.1 of Chapter IV, we know that $\mathcal{I}_1 \approx \mathcal{O}_{Y_1}(1)$ and is invertible. Tensoring with $\mathcal{O}_{Y_1}(n)$ the exact sequence

$$0 \to \mathcal{I}_1 \to \mathcal{O}_{Y_1} \to f_{1*} \mathcal{O}_{X_1} \to 0.$$

yields the exact sequence

$$0 \to \mathcal{O}_{Y_1}(n+1) \to \mathcal{O}_{Y_1}(n) \to f_{1*}(\mathcal{O}_{X_1}(n)) \to 0.$$

We apply the functor φ_*. We note that $\varphi_* f_{1*} = f_* \psi_*$. Furthermore f, f_1 are closed imbeddings, so $R^i f_* = 0$ and $R^i f_{1*} = 0$ for $i \geqq 1$. Then we get:

$$\begin{aligned} R^i \varphi_*(f_{1*} \mathcal{O}_{X_1}(n)) &= R^i(\varphi_* f_{1*})(\mathcal{O}_{X_1}(n)) \\ &= R^i(f_* \psi_*)(\mathcal{O}_{X_1}(n)) \\ &= f_*(R^i \psi_*)(\mathcal{O}_{X_1}(n)) = \begin{cases} f_*(\mathrm{Sym}^n \mathscr{C}_{X/Y}) & \text{if } i = 0, \\ 0 & \text{if } i \geqq 1 \end{cases} \end{aligned}$$

by the fundamental properties **R 5** and **R 6** of the cohomology, Chapter V, §2. The long cohomology sequence then yields an isomorphism

$$0 \to R^i \varphi_* \mathcal{O}_{Y_1}(n+1) \xrightarrow{\approx} R^i \varphi_* \mathcal{O}_{Y_1}(n) \to 0$$

for $i \geqq 1$ and all $n \geqq 0$. By **R 4** (Serre's theorem), $R^i \varphi_* \mathcal{O}_{Y_1}(n) = 0$ for n sufficiently large, so $= 0$ for all $n \geqq 0$. Thus

$$R^i \varphi_* \mathcal{O}_{Y_1} = 0 \qquad \text{for} \quad i \geqq 1.$$

Now let $i = 0$ and $n \geqq 0$. We get an exact sequence

$$0 \to \varphi_* \mathcal{O}_{Y_1}(n+1) \to \varphi_* \mathcal{O}_{Y_1}(n) \to f_* \operatorname{Sym}^n(\mathscr{C}) \to 0.$$

Since $Y_1 = \operatorname{Proj}\left(\bigoplus_{n=0}^{\infty} \mathscr{I}^n\right)$ there exist canonical homomorphisms of sheaves

$$\mathscr{I}^n \to f_* \mathcal{O}_{Y_1}(n)$$

giving rise to the commutative diagram

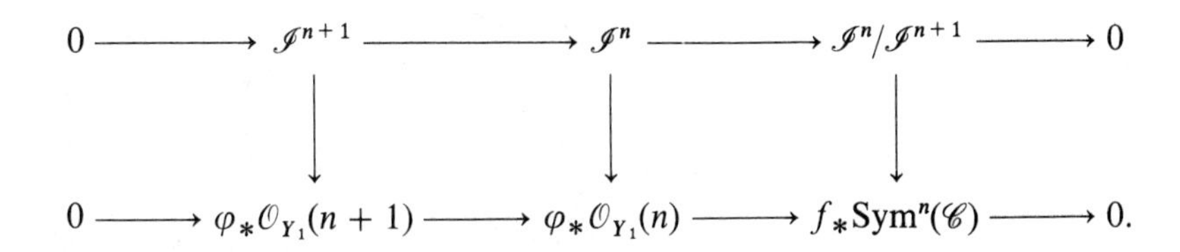

The left vertical arrow is an isomorphism for all sufficiently large n by Serre's theorem. The right vertical arrow is an isomorphism by Corollary 2.4 of Chapter IV. By descending induction on n it follows that the center arrow is an isomorphism for all $n \geqq 0$. This concludes the proof of Proposition 4.1.

Proposition 4.1 was the last geomeric fact needed to determine most of the structure of K of a blow up, and all of it in the case when the schemes are regular. We shall now enter into formal considerations, so we make a precise list of what we use.

Let K be a λ-ring functor. Let $f: X \to Y$ be a morphism. We say that f satisfies the **self intersection formula with multiplier** $\lambda_{-1}(c)$ for some $c \in K(X)$ if

$$f^K f_K(x) = \lambda_{-1}(c)x \qquad \text{for all} \quad x \in K(X).$$

Consider a commutative diagram:

$$\begin{array}{ccc} X_1 & \xrightarrow{f_1} & Y_1 \\ {\scriptstyle\psi}\downarrow & & \downarrow{\scriptstyle\varphi} \\ X & \xrightarrow[f]{} & Y \end{array}$$

We say that this diagram is a **blow up diagram with respect to** K if the following conditions are satisfied:

Bl 1. ψ is an elementary projection with respect to K, in the sense of Chapter II, §2.

We recall what this means: $K(X_1)$ as $K(X)$-algebra via ψ^K is isomorphic with the extension $K(X)_c$ of $K(X)$ for some positive element $c \in K(X)$ (cf. Chapter I, §2), and ψ_K corresponds to the associated functional ψ_c. We let

$$K(X)_c = K(X)[\ell],$$

where ℓ is the canonical generator.

Bl 2. $\varphi_K(1) = 1$ and therefore $\varphi_K \varphi^K \colon K(Y) \to K(Y)$ is the identity (by the projection formula).

Bl 3. f and f_1 satisfy the self intersection formula with multipliers $\lambda_{-1}(c)$ and $\lambda_{-1}(\ell)$ respectively.

Bl 4. Let $e = c - \ell$, or more precisely $e = \psi^K(c) - \ell$. Then the diagram satisfies the Intersection Formula

$$\varphi^K f_K(x) = f_{1K}(\lambda_{-1}(e)\psi^K(x))$$

for all $x \in K(X)$.

We have proved that the blow up diagram arising from blowing up a regular imbedding in the category of schemes satisfies the **Bl** properties: **Bl 1** comes from Chapter IV, Lemma 4.1; **Bl 2** is Proposition 4.1; **Bl 3** comes from Theorem 1.3, special case 1.3.2; and **Bl 4** is once more the Intersection Formula of Theorem 1.3, special case 1.3.3. We now work only with these properties, unless otherwise specified.

Lemma 4.2. *Let ψ be as in* **Bl 1**, *and let f_1 satisfy the self intersection formula with multiplier $\lambda_{-1}(\ell)$ as in* **Bl 3**. *Let $e = \psi^K(c) - \ell$. If $x_1 \in \operatorname{Ker} f_{1K}$, then*

$$x_1 = \lambda_{-1}(e)\psi^K \psi_K(x_1).$$

Proof. By the self intersection formula for f_1 we have

$$0 = f_1^K f_{1K}(x_1) = \lambda_{-1}(\ell)x_1 = (1 - \ell)x_1.$$

Hence the lemma results from the next lemma on λ-rings, cf. [SGA 6], VI, Proposition 5.10.

Lemma 4.3. *Let K be λ-ring. Let c be a positive element in K and let $K_c = K[\ell]$ be the extension of Chapter* I, §2 *with associated functional $\psi_c\colon K_c \to K$. Let $z \in K_c$ and assume that $z(1 - \ell) = 0$. Then*

$$z = \psi_c(z)\lambda_{-1}(c - \ell).$$

Proof. Let $\varepsilon(c) = r + 1$. Write z as a linear combination

$$z = \sum_{i=0}^{r} a_i(\ell - 1)^i \qquad \text{with} \quad a_i \in K.$$

From $z(\ell - 1) = 0$ we get

$$0 = \sum_{i=1}^{r+1} a_{i-1}(\ell - 1)^i.$$

On the other hand, by Proposition 1.1(a) of Chapter III, with $t = 1 - \ell$, we know that the equation for ℓ over K can also be written

$$0 = \sum_{i=0}^{r+1} (-1)^i \gamma^{r+1-i}(c - r - 1)(\ell - 1)^i.$$

Multiplying this equation by a_r and comparing coefficients show that

$$a_r(-1)^{r+1-i}\gamma^{r+1-i}(c - r - 1) = a_{i-1} \qquad \text{for} \quad i = 1, \ldots, r + 1.$$

Hence

$$\begin{aligned}
z = \sum_{i=1}^{r+1} a_{i-1}(\ell - 1)^{i-1} &= a_r \sum_{i=1}^{r+1} (-1)^{r+1-i}\gamma^{r+1-i}(c - r - 1)(\ell - 1)^{i-1} \\
&= a_r(-1)^r \sum_{i=0}^{r} \gamma^{r-i}(c - r - 1)\gamma^i(1 - \ell) \\
&= a_r(-1)^r\gamma^r(c - r - \ell) \qquad \text{because } \gamma \text{ is a } \lambda\text{-operation} \\
&= a_r\lambda_{-1}(c - \ell)
\end{aligned}$$

by putting $t = 0$ in Proposition 1.1(a) of Chapter III. Applying ψ_c and using Corollary 2.3 of Chapter I with $t = -1$ yields

$$\psi_c(z) = a_r.$$

This concludes the proof of Lemma 4.3, and hence of Lemma 4.2.

We come to the desired exact sequence for K of a blow up. The following theorem axiomatizes [SGA 6], VII, Theorem 3.7.

Theorem 4.4. *Let*

$$\begin{array}{ccc} X_1 & \xrightarrow{f_1} & Y_1 \\ \psi \downarrow & & \downarrow \varphi \\ X & \xrightarrow[f]{} & Y \end{array}$$

be a blow up diagram, and let $e = \psi^K(c) - \ell$. *Then the following sequence is exact:*

$$0 \longrightarrow K(X) \xrightarrow{u} K(X_1) \oplus K(Y) \xrightarrow{v} K(Y_1)$$

where u, v *are the homomorphisms defined by:*

$$u(x) = (-\lambda_{-1}(e)\psi^K(x), f_K(x))$$
$$v(x_1, y) = f_{1K}(x_1) + \varphi^K(y).$$

The sequence splits with the left inverse u' *for* u *given by*

$$u'(x_1, y) = -\psi_K(x_1),$$

that is $u'u = \mathrm{id}_{K(X)}$.

Proof. We proceed stepwise.

u is injective, split by u′. Indeed, by the projection formula

$$u'u(x) = \psi_K(\lambda_{-1}(e)\psi^K(x)) = \psi_K(\lambda_{-1}(e))x = x$$

by Corollary 2.3 of Chapter I, with $t = -1$.

$\mathbf{v \circ u = 0}$ is just the Intersection Formula **Bl 4**, because

$$v(u(x)) = \varphi^K f_K(x) - f_{1K}(\lambda_{-1}(e)\psi^K x).$$

Ker v ⊂ Im u. Since u' splits u we have a direct sum decomposition

$$K(X_1) \oplus K(Y) = \mathrm{Im}\, u \oplus \mathrm{Ker}\, u',$$

where directly from the definition,

$$\operatorname{Ker} u' = \operatorname{Ker} \psi_K \oplus K(Y).$$

Let $(x_1, y) \in \operatorname{Ker} u'$ so that $\psi_K(x_1) = 0$, and suppose $v(x_1, y) = 0$, that is

$$f_{1K}(x_1) + \varphi^K(y) = 0.$$

Applying φ_K and using **Bl 2** yields

$$0 = \varphi_K f_{1K}(x_1) + \varphi_K \varphi^K(y) = f_K \psi_K(x_1) + y = y.$$

Thus $y = 0$. Then $f_{1K}(x_1) = 0$ and $x_1 = 0$ by assumption and Lemma 4.2. This concludes the proof that the sequence is exact.

Remark. If, as in the next result, v is surjective, then v gives an *additive* isomorphism

$$v\colon \operatorname{Ker} \psi_K \oplus K(Y) \xrightarrow{\approx} K(Y_1).$$

The result depends on more than the formal **Bl** conditions.

Theorem 4.5. *In the blow up diagram as at the beginning of the section, suppose that X and Y are regular schemes, so X_1, Y_1 are also regular. Then v is surjective, and hence we have the exact sequence*

$$0 \longrightarrow K(X) \xrightarrow{u} K(X_1) \oplus K(Y) \xrightarrow{v} K(Y_1) \longrightarrow 0.$$

Proof. Under the regularity assumption and Proposition 3.1 we can identify $K^{\cdot} = K = K_{.}$ so we can use the exact sequence of Proposition 3.2, which yields in the present instance exactly

$$K(X_1) \xrightarrow{f_{1K}} K(Y_1) \xrightarrow{j_1^K} K(Y_1 - X_1) \longrightarrow 0,$$

where $j_1\colon Y_1 - X_1 \to Y_1$ is the inclusion, and similarly

$$K(Y) \xrightarrow{j^K} K(Y - X) \longrightarrow 0.$$

Furthermore, φ induces an isomorphism $\varphi_{Y-X}\colon Y_1 - X_1 \to Y - X$. Hence given $y_1 \in K(Y_1)$ there exists $y \in K(Y)$ such that

$$j_1^K y_1 = \varphi_{Y-X}^K j^K(y) = j_1^K \varphi^K y.$$

Hence $j_1^K(y_1 - \varphi^K y) = 0$ so there exists $x_1 \in K(X_1)$ such that

$$y_1 - \varphi^K y = f_{1K} x_1.$$

This proves that v is surjective, and concludes the proof of the theorem.

The rest of the section goes back to the formal **Bl** conditions.

Proposition 4.6. *Let $f: X \to Y$ be a morphism satisfying the self intersection formula with multiplier $\lambda_{-1}(c)$. Then*

$$f_K(\lambda_{-1}(c)xx') = f_K(x)f_K(x').$$

Proof. The self intersection formula reads

$$f^K f_K(x) = \lambda_{-1}(c)x.$$

Then

$$f_K(\lambda_{-1}(c)xx') = f_K(f^K f_K(x)x') = f_K(x)f_K(x')$$

by the projection formula. This proves the proposition.

The above proposition suggests redefining a product in $K(X)$ in such a way that f_K *becomes a multiplicative homomorphism*, namely we **define**

$$x * x' = \lambda_{-1}(c)xx'.$$

This product is associative and commutative, and makes $K(X)$ into an algebra (**Z**-algebra), even into a $K(Y)$-algebra via f^K as one immediately verifies using the projection formula. Note however, that this star multiplication does not necessarily have a unit element.

Similarly, we redefine the multiplication in $K(X_1)$ by

$$x_1 * x_1' = \lambda_{-1}(\ell)x_1 x_1' = (1 - \ell)x_1 x_1'.$$

Then f_{1K} is a multiplicative homomorphism for this star multiplication.

Warning. Even though we are used to imbedding $K(X)$ in $K(X_1)$ via ψ^K, the multiplication we have just defined in $K(X_1)$ does not induce the star multiplication in $K(X)$. Indeed if we identify $K(X)$ in $K(X_1)$ then we have

$$e = c - \ell,$$

and since λ_{-t} is a homomorphism, we get

$$\lambda_{-1}(c) = \lambda_{-1}(e)\lambda_{-1}(\ell) = \lambda_{-1}(e)(1 - \ell).$$

The star multiplication in $K(X_1)$ could be denoted more accurately by

$$x_1 *_1 x'_1$$

but for simplicity of notation, we shall omit the index on this star.

The groups $K(X)$ and $K(X_1)$ with the star multiplication will be denoted by

$$K(X)_* \qquad \text{and} \qquad K(X_1)_*$$

respectively.

We introduce a **star multiplication** on the direct sum $K(X_1)_* \oplus K(Y)$ by defining

$$(x_1, y)(x'_1, y') = (z_1, yy'),$$

where

$$z_1 = \lambda_{-1}(\ell)x_1x'_1 + x'_1\psi^K f^K y + x_1\psi^K f^K y'.$$

This makes the direct sum into a commutative algebra. Note that the summands $K(X_1)_*$ and $K(Y)$ have the star and ordinary multiplications in their natural imbedding in the direct sum.

Theorem 4.7. *With the star multiplications in $K(X)_*$ and $K(X_1)_*$ and the above multiplication on the direct sum:*

(i) *u and v are multiplicative homomorphisms, and u is a homomorphism of $K(Y)$-algebras.*
(ii) Im *u is an ideal in $K(X_1)_* \oplus K(Y)$, and in fact*

$$(x_1, y)u(x) = u(f^K(y)x).$$

(iii) Im *u and $K(X_1)_*$ are orthogonal with respect to the multiplication in $K(X_1)_* \oplus K(Y)$.*

Proof. That u is a homomorphism follows at once from the definitions and Proposition 4.6, using

$$\lambda_{-1}(c) = \lambda_{-1}(e)\lambda_{-1}(\ell).$$

There is also no difficulty in verifying that u is a homomorphism of $K(Y)$-algebras. Similarly one verifies that v is a homomorphism. We write out in full the proof of (ii). We have:

$$(x_1, y)u(x) = (x_1, y)(-\lambda_{-1}(e)\psi^K x, f_K x) = (z_1, yf_K(x)),$$

where

$$z_1 = -\lambda_{-1}(\ell)\lambda_{-1}(e)x_1\psi^K x + x_1\psi^K f^K f_K x - \lambda_{-1}(e)\psi^K x\psi^K f^K y.$$

By the self intersection formula $f^K f_K x = \lambda_{-1}(c)x$ of **Bl 3**, and the projection formula we get

$$(x_1, y)u(x) = u(f^K(y)x).$$

This shows both that the image of u is an ideal, and also that the image of u is orthogonal to $K(X_1)_*$ (when $y = 0$), thereby concluding the proof of the theorem.

VI §5. Upper and Lower Filtrations

In this section we work with the same category $\mathfrak{C}$ as in §3, that is a category of Noetherian schemes each of which has an ample invertible sheaf. The morphisms are arbitrary scheme morphisms.

We discuss a filtration on $K.X$ compatible with the filtration of $K^{\cdot}X$ defined in Chapter V, §1. We define the **lower filtration**:

$F_m K.X$ = set of elements $x \in K.X$ such that there exist coherent sheaves $\mathcal{F}_1$, $\mathcal{F}_2$ satisfying

$$x = [\mathcal{F}_1] - [\mathcal{F}_2] \quad \text{and} \quad \dim \operatorname{Supp}(\mathcal{F}_i) \leqq m, \quad i = 1, 2.$$

Proposition 5.1. *The subgroup $F_m K.X$ is generated by the classes $[\mathcal{O}_V]$, where V runs through the integral closed subschemes of X of dimension at most m.*

Proof. It suffices to prove that for a coherent sheaf $\mathcal{F}$ with

$$\dim \operatorname{Supp}(\mathcal{F}) \leqq m,$$

we have

$$[\mathcal{F}] = \sum \ell_V(\mathcal{F})[\mathcal{O}_V] \mod F_{m-1}K.K, \tag{5.1}$$

where the sum is over the m-dimensional irreducible components V of $\mathrm{Supp}(\mathscr{F})$, and $\ell_V(\mathscr{F})$ denotes the length of the stalk of $\mathscr{F}$ at the generic point of V. For sheaves $\mathscr{F}$ with support contained in a given closed subset Z of dimension at most m, both sides of (5.1) are exact, so one may induct on $\ell_V(\mathscr{F})$. If $\mathscr{I}$ is the ideal sheaf of a component V of Z, the exact sequence

$$0 \to \mathscr{I}^{n-1}\mathscr{F}/\mathscr{I}^n\mathscr{F} \to \mathscr{F}/\mathscr{I}^n\mathscr{F} \to \mathscr{F}/\mathscr{I}^{n-1}\mathscr{F} \to 0$$

and the fact that $\mathscr{I}^n\mathscr{F} = 0$ for n large shows that we may assume $V = Z$ and $\mathscr{F}$ is a coherent sheaf of $\mathscr{O}_V$-modules. If $r = \ell_V(\mathscr{F})$, there is a non-empty open set U of V and an isomorphism of $\mathscr{O}_U^{\oplus r}$ with $\mathscr{F}|U$. By Lemma 3.7 there is a coherent sheaf $\mathscr{G}$ on V and homomorphism

$$\mathscr{G} \to \mathscr{O}_V^{\oplus r} \quad \text{and} \quad \mathscr{G} \to \mathscr{F}$$

which are isomorphisms over U. Since the kernel and cokernels of these homorphisms define classes in $F_{m-1}K.X$, it follows that

$$[\mathscr{F}] = [\mathscr{O}_V^{\oplus r}] = \ell_V(\mathscr{F})[\mathscr{O}_V] \mod F_{m-1}K.X,$$

as required.

Proposition 5.2. *Under the product $K^{\cdot}X \otimes K.X \to K.X$, we have the inclusion*

$$F^nK^{\cdot}(X) \cdot F_mK.X \subset F_{m-n}K.X.$$

Proof. We show in fact that

$$F^n_{\mathrm{top}}K^{\cdot}X \cdot F_mK.X \subset F_{m-n}K.X,$$

which is stronger by Chapter V, Theorem 3.9. Given $x \in F^n_{\mathrm{top}}K^{\cdot}X$, $y \in F_mK.X$, we may assume $y = [\mathscr{F}]$ for a coherent sheaf $\mathscr{F}$ whose support Y has dimension at most m. Then x is represented by a complex $\mathscr{E}^{\cdot}$ of locally free sheaves which is exact off a closed subset Z of X with

$$\mathrm{codim}(Z \cap Y, Y) \geqq n;$$

therefore $\dim(Z \cap Y) \leqq m - n$. Then

$$x \cdot y = \sum (-1)^i[\mathscr{E}^i \otimes \mathscr{F}] = \sum (-1)^i[\mathscr{H}^i(\mathscr{E}^{\cdot} \otimes \mathscr{F})],$$

and $\mathrm{Supp}(\mathscr{H}^i(\mathscr{E}^{\cdot} \otimes \mathscr{F})) \subset \mathrm{Supp}(\mathscr{H}^i\mathscr{E}) \cap \mathrm{Supp}(\mathscr{F}) \subset Z \cap Y$, which proves that $x \cdot y$ as in $F_{m-n}K.X$, and concludes the proof of the proposition.

We recall that $\delta: K^{\cdot}X \to K_{\cdot}X$ was the natural homomorphism induced by the inclusion of the category of locally free sheaves into the category of coherent sheaves.

Proposition 5.3. *Let d be the dimension of X. Then*

$$\delta(F^n K^{\cdot}X) \subset F_{d-n} K_{\cdot}X.$$

Proof. This follows immediately from Proposition 5.2.

Let $G^n X = \mathrm{Gr}^n K^{\cdot}X = F^n K^{\cdot}X / F^{n+1} K^{\cdot}X$ be the associated graded group studied in Chapters III and V, and set

$$G^{\cdot}X = \bigoplus_{n \geqq 0} G^n X = \bigoplus_{n \geqq 0} \mathrm{Gr}^n K^{\cdot}X$$

Define the **lower graded component**

$$G_m X = \mathrm{Gr}_m K_{\cdot}X = F_m K_{\cdot}X / F_{m-1} K_{\cdot}X,$$

and set

$$G_{\cdot}X = \bigoplus_{m \geqq 0} G_m X = \bigoplus_{m \geqq 0} \mathrm{Gr}_m K_{\cdot}X.$$

By Proposition 5.2, tensor product induces a **"cap" product**

$$G^n X \otimes G_m X \xrightarrow{\ \cap\ } G_{m-n} X,$$

making $G_{\cdot}X$ into a graded $G^{\cdot}X$-module. (The notation "$\cap$" is to suggest the cap product of topology.)

By Proposition 5.3 we conclude that δ induces a homomorphism on the graded groups

$$\delta_G : \mathrm{Gr}\, K^{\cdot}(X) \to \mathrm{Gr}\, K_{\cdot}(X)$$

such that $\delta_G(x) = x \cap [\mathcal{O}_X]$. Actually we have an induced map on each graded component

$$\delta_G : \mathrm{Gr}^n K^{\cdot}(X) \to \mathrm{Gr}_{d-n} K_{\cdot}(X).$$

The commutativity relation of Proposition 3.3 for δ now gives the corresponding relation in the graded context:

Corollary 5.4. *Let* $G = \mathbf{Q}\mathrm{Gr}\, K$. *For any regular morphism* $f: X \to Y$, *the following diagram commutes.*

$$\begin{array}{ccc} G^{\cdot}(X) & \xrightarrow{\delta_G} & G_{.}(X) \\ \downarrow{\scriptstyle f_{G^{\cdot}}} & & \downarrow{\scriptstyle f_{G_{.}}} \\ G^{\cdot}(Y) & \xrightarrow[\delta_G]{} & G_{.}(Y) \end{array}$$

Example. Let k be a field and $Y = \mathrm{Spec}(k)$. Let $f: X \to Y$ be a regular morphism and let $d = \dim X$. By Proposition 5.1, $\mathbf{Q}\mathrm{Gr}_0\, K_{.}(X) = G_0(X)$ is generated by the classes $[\mathcal{O}_P]$, where P ranges over the closed points. In this case, using the functoriality on the composite $P \to X \to Y$, one sees at once that

$$f_{G_{.}}[\mathcal{O}_P] = [k(P): k][\mathcal{O}_Y] = [k(P): k],$$

where we identify $G_{.}(Y)$ with $\mathbf{Q}$ via the basis element $[\mathcal{O}_Y]$.

With this identification the commutative diagram of Corollary 5.4 on the component of top degree reads:

$$\begin{array}{ccc} \mathbf{Q}\mathrm{Gr}^{\mathrm{top}}\, K^{\cdot}(X) & \xrightarrow{\delta_G} & \mathbf{Q}\mathrm{Gr}_0\, K_{.}(X) \\ & {\scriptstyle f_G = f_{G^{\cdot}}} \searrow \quad \swarrow {\scriptstyle f_{G_{.}}} & \\ & \mathbf{Q} & \end{array}$$

This gives the promised geometric interpretation of f_G in top graded degree, relevant for the complete interpretation of the Hirzebruch Riemann–Roch theorem of Chapter V, Corollary 7.4. Indeed, $f_{G_{.}}$ is the ordinary "degree" of 0-cycles, in which case we have a preconceived geometric notion of "number of points".

In the preceding chapter, we compared the γ-filtration $F^n K(X)$ with a topological filtration $F^n_{\mathrm{top}} K(X)$. There is another natural filtration of $K(X)$ when X is regular. We let:

$'F^n_{\mathrm{top}} K(X) =$ subgroup of $K(X)$ generated by classes $[\mathcal{F}]$ of coherent sheaves $\mathcal{F}$ whose supports have codimension at least n in X.

As in Proposition 5.1, ${}'F^n_{\text{top}}K(X)$ is generated by the classes $[\mathcal{O}_V]$, where V runs through the integral closed subschemes of X of codimension at least n. Note that in case X has dimension d, and

$$\dim(V) + \operatorname{codim}(V, X) = d$$

for all such V (for example, if X is a variety over a field), then

$${}'F^n_{\text{top}}K(X) = F_{d-n}K(X).$$

On a general scheme, however, one must distinguish these notions.

Proposition 5.5. *If X is regular, then*

$$F^nK(X) \subset F^n_{\text{top}}K(X) \subset {}'F^n_{\text{top}}K(X),$$

and

$$\mathbf{Q}F^nK(X) = \mathbf{Q}F^n_{\text{top}}K(X) = \mathbf{Q}'F^n_{\text{top}}K(X)$$

in $\mathbf{Q}K(X)$.

Proof. The first inclusion was proved in Chapter V, Theorem 3.9. The second follows from the fact that if $\mathcal{E}^\cdot$ is any bounded complex of locally free (or coherent) sheaves, with homology sheaves $\mathcal{H}^i$, then

$$\sum (-1)^i[\mathcal{E}^i] = \sum (-1)^i[\mathcal{H}^i]$$

in $K(X)$.

To show that all three agree after tensoring with $\mathbf{Q}$, we must show that if V is an integral closed subscheme of X, and $n = \operatorname{codim}(V, X)$, then

$$[\mathcal{O}_V] \in \mathbf{Q}F^nK(X).$$

By Noetherian induction, we may assume this has been proved for all proper closed integral subschemes of V.

There is a proper closed subscheme S of V such that the inclusion

$$j: V - S \to X - S$$

is a regular imbedding of codimension n. One sees this by taking n equations which generate the ideal of V in the local ring of X at the generic point of V; such a sequence is regular on some open set U, and one may choose S so that its complement in V is $V \cap U$.

Since j is a regular imbedding, we have seen that

$$j_K(F^k K(V - S)) \subset \mathbf{Q}F^{k+n}K(X - S)$$

(Chapter V, Proposition 6.4). In particular,

$$[\mathcal{O}_{V-S}] \in \mathbf{Q}F^n K(X - S).$$

Consider the exact sequence of Proposition 3.2.

$$\mathbf{Q}K(S) \to \mathbf{Q}K(X) \to \mathbf{Q}K(X - S) \to 0.$$

Since the restriction map $K(X) \to K(X - S)$ is a surjection of λ-rings, it maps $F^n K(X)$ onto $F^n K(X - S)$. Therefore there is an x in $\mathbf{Q}F^n K(X)$ such that

$$y = [\mathcal{O}_V] - x \in \operatorname{Im}(\mathbf{Q}K(S) \to \mathbf{Q}K(\mathrm{X})).$$

Expressing y as a rational combination of classes $[\mathcal{O}_W]$, for W integral closed subschemes of S, and applying Noetherian induction to these W, gives

$$[\mathcal{O}_V] = x + y \in \mathbf{Q}F^n K(X),$$

as required.

Finally we deal with the functoriality of $f_{K.}$ with respect to the filtration.

Proposition 5.6. *If $f\colon X \to Y$ is proper then*

$$f_{K.}(F_m K.X) \subset F_m K.Y.$$

Proof. It suffices to note that

$$\operatorname{Supp}(R^i f_* \mathscr{F}) \subset f(\operatorname{Supp} \mathscr{F}),$$

and $\dim f(Z) \leqq \dim Z$ for any Z closed in X.

The lemma tells us that $f_{K.}$ is compatible with the lower filtration. Therefore we have an induced functorial homomorphism

$$f_{G.}\colon G.(X) \to G.(Y)$$

for a proper morphism f.

By the projection formula of §3 for $K^{\cdot}$ and $K_{\cdot}$ and Proposition 5.6 we deduce the **projection formula for the graded functors**

$$f_{G_{\cdot}}(f^{G^{\cdot}}(y) \cap x) = y \cap f_{G_{\cdot}}(x)$$

for $f: X \to Y$ proper, $x \in G_{\cdot}X$ and $y \in G^{\cdot}Y$.

VI §6. The Contravariant Maps $f^{K_{\cdot}}$ and $f^{G_{\cdot}}$.

From here on, we only give indications of proofs, if at all.

We have defined $K_{\cdot}$ as a covariant functor for proper morphisms. We now wish to define $K_{\cdot}$ as a contravariant functor. In order to take care of the open subschemes as in §3, we let:

$\mathfrak{C}$ = category whose objects are the same as in §3 and whose morphisms are those which can be factored as $p \circ i$, where p is smooth and i is a regular imbedding.

Note here that the only difference with our previous regular morphisms is that p is not assumed proper. We assume now that morphisms are in this category.

Suppose first that $f: X \to Y$ is flat. Then the obvious desideratum gives us the contravariant map, namely for $\mathcal{G}$ coherent on Y,

$$f^{K_{\cdot}}[\mathcal{G}] = [f^*\mathcal{G}].$$

This does give a homomorphism $K_{\cdot}(Y) \to K_{\cdot}(X)$ since f^* is exact. If f is smooth, then f is flat, and this definition applies.

If f is not flat, there is a technical complication, and we have to go through the same rigamarole as before, which we summarize. Let us begin by a sheaf-theoretic remark. Let X be a closed subscheme of Y. Let $\mathcal{H}$ be a coherent sheaf on Y, supported by X. Let $\mathcal{I}$ be the ideal sheaf defining X in $\mathcal{O}_Y$. Then there is some power $\mathcal{I}^m$ such that

$$\mathcal{I}^m\mathcal{H} = 0.$$

Therefore there is a filtration

$$\mathcal{H} \supset \mathcal{I}\mathcal{H} \supset \mathcal{I}^2\mathcal{H} \supset \cdots \supset \mathcal{I}^m\mathcal{H} = 0$$

such that each factor sheaf $\mathcal{I}^r\mathcal{H}/\mathcal{I}^{r+1}\mathcal{H}$ is a coherent sheaf over $\mathcal{O}_X$. We define

$$[\mathcal{H}]_X = \sum_{r=0}^{m-1} [\mathcal{I}^r\mathcal{H}/\mathcal{I}^{r+1}\mathcal{H}]_X,$$

where the subscript X indicates the class in $K.(X)$, which is defined for each term on the right-hand side, and thus defines the left-hand side.

Lemma 6.1. *Let $i: X \to Y$ be a closed imbedding. Let $K.(Y, X)$ be the Grothendieck group of coherent sheaves on Y supported by X. The homomorphism*

$$i_{K.}: K.(X) \to K.(Y, X)$$

induced by i_* is an isomorphism, whose inverse is given by

$$[\mathcal{H}]_Y \mapsto [\mathcal{H}]_X$$

as defined above.

Proof. This is an easy consequence of the Jordan–Hölder theorem, which we leave to the reader.

Note. For clarity we indexed the class of a sheaf by Y and X respectively. In practice, we may also drop the indices by making the identification via the isomorphism of the lemma, or we may just write the X as an index to make the distinction clear.

Suppose that $i: X \to Y$ is a regular imbedding. There is a finite resolution

$$\mathcal{E}. \to \mathcal{O}_X \to 0$$

by locally free sheaves on Y (e.g. the Koszul complex). For any coherent sheaf $\mathcal{G}$ on Y we then obtain a complex $\mathcal{E}. \otimes \mathcal{G}$. We define

$$i^{K.}[\mathcal{G}] = \sum_k (-1)^k [\mathcal{H}_k(\mathcal{E}. \otimes \mathcal{G})]_X,$$

where $\mathcal{H}_k(\mathcal{E}. \otimes \mathcal{G})$ is the k-th homology of $\mathcal{E}. \otimes \mathcal{G}$, and is supported by X. By basic abstract nonsense of homological algebra (sheaf Tor), the sheaf homology is independent of the resolution. Since in addition $\mathcal{G} \mapsto \mathcal{E}. \otimes \mathcal{G}$ is exact, we obtain a well-defined map

$$i^{K.}: K.(Y) \to K.(X).$$

If $f: X \to Y$ is factored as $f = p \circ i$, then by using the same steps as in the definition of f^K in Chapter V, §5 one can prove that the map

$$i^{K.} \circ p^{K.}$$

is independent of the factorization, and defines $f^{K.}$ functorially. It can also be verified that for flat f, the map $f^{K.}$ obtained from a factorization is the pull-back that we mentioned first.

Having defined $K.$ as contravariant functor, it is then natural to obtain the corresponding **Projection Formula**:

Proposition 6.2. *Let $f: X \to Y$ be a regular morphism. For $x \in K^{\cdot}(X)$ and $y \in K.(Y)$ we have*

$$f_{K.}(x \cdot f^{K.}(y)) = f_{K^{\cdot}}(x) \cdot y.$$

Proof. By factoring f into a regular imbedding and a projection, it suffices to prove the formula in each case. In the case of a projection, the formula follows from **R 7** of Chapter V, §2 just as the first projection formula of Proposition 3.4 followed from **R 3** after we use the fact that the classes of sheaves $[\mathcal{O}(n)]$ generate $K^{\cdot}\mathbf{P}$ over the base for a projective bundle $\mathbf{P}$.

In the case of a regular imbedding, the formula involves two resolutions, and a proof can be given by constructing a double complex, in the style of general homological algebra.

The next results have to do with the graded properties of $f^{K.}$, and so involve dimension as well as codimension. This means that one has to be careful about the schemes involved. Therefore we assume in addition, for the rest of this section, that

all schemes are over a field.

Schemes of finite type over a regular base would suffice, provided that an appropriate notion of dimension is used. See [F 2], Chapter 20.

Next we pass to the grading properties of $f^{K.}$.

Proposition 6.3. *Let $f: X \to Y$ be a morphism in $\mathfrak{C}$, of codimension d. Then*

$$f^{K.}(F_m K.Y) \subset F_{m-d}K.X$$

so $f^{K.}$ induces a functorial homomorphism

$$f^{G.}: G_m(Y) \to G_{m-d}(X).$$

Proof. It suffices to prove the proposition when f is smooth and when f is a regular imbedding. In the first case, one uses Proposition 5.1, and the assertion is immediate by applying $f^{K\cdot}$ to $[\mathcal{O}_V]$ where V has dimension m. In the case of a regular imbedding, a proof can be given by deformation to the normal bundle. Note that the proof of Chapter V, Proposition 6.4 that $f_{K.}$ has a graded degree also went through deformation to the normal bundle, via Adams Riemann–Roch.

Then we have the projection formula for the graded map:

Proposition 6.4. *Let $f: X \to Y$ be a regular morphism. Let $G = \mathbf{Q}\mathrm{Gr}\, K$. For $x \in G^{\cdot}(X)$ and $y \in G.(Y)$ we have*

$$f_{G.}(x \cap f^{G\cdot}(y)) = f_{G^{\cdot}}(x) \cap y.$$

Proof. This is an immediate consequence of the non-graded Proposition 6.2 together with the compatibility with filtrations and the induced graded maps which has been proved in all cases.

In the next section, we shall state a Riemann–Roch theorem involving the contravariant maps introduced above.

A particular case of Proposition 6.3 occurs for the restriction

$$f^{K\cdot}: K.Y \to K.U$$

to an open subscheme of Y, and the induced map

$$f^{G\cdot}: G_m(Y) \to G_m(U).$$

Proposition 6.5. *If U is the complement of a closed subscheme X of Y, then we have an exact sequence*

$$G_m(X) \to G_m(Y) \to G_m(U) \to 0$$

Although this proposition looks innocuous, and is the graded analogue of Proposition 3.2, we don't know any proof which does not involve using the Singular Riemann–Roch theorem with values in the Chow group, of [BFM 1], which we discuss in the next section.

VI §7. Singular Riemann–Roch

This section uses only §3, §5, *and* §6.
We continue with the same category and with the same notation.

The Riemann–Roch theorem of Chapter V, §6 yields a formula for the Euler characteristic $\chi(X, \mathscr{E})$ of a locally free sheaf $\mathscr{E}$ on a projective scheme X over a field k only if X is a local complete intersection, i.e. the morphism from X to Spec(k) is regular in the sense of Chapter V, §5.

We next give a statement of a Riemann–Roch theorem for more singular schemes. *For simplicity we restrict our attention to schemes which are quasi-projective over a field k.* Much of the theorem is valid without this assumption. The main use of a ground field is to have dimensions and codimensions of closed and open subschemes behave nicely. For more general versions see [F 2], 20. We set $S = \mathrm{Spec}(k)$.

The **singular Riemann–Roch theorem** constructs a homomorphism

$$\tau = \tau_X : K.X \to \mathbf{Q}G.X$$

satisfying the following properties:

SRR 1. **(Covariance).** *If* $f: X \to Y$ *is proper, then the following diagram commutes.*

$$\begin{array}{ccc} K.X & \xrightarrow{\tau} & \mathbf{Q}G.X \\ \downarrow{\scriptstyle f_{K.}} & & \downarrow{\scriptstyle f_{G.}} \\ K.Y & \xrightarrow{\tau} & \mathbf{Q}G.Y \end{array}$$

SRR 2. **(Module).** *For any* X, *the following diagram commutes.*

$$\begin{array}{ccc} K^{\cdot}X \otimes K.X & \xrightarrow{\mathrm{ch} \otimes \tau} & \mathbf{Q}G^{\cdot}X \otimes \mathbf{Q}G.X \\ \downarrow{\scriptstyle \cdot} & & \downarrow{\scriptstyle \cap} \\ K.X & \xrightarrow{\tau} & \mathbf{Q}G.X \end{array}$$

For any X one then defines the **Todd class** $\mathrm{Td}(X)$ in $\mathbf{Q}G.X$ by

$$\mathrm{Td}(X) = \tau([\mathcal{O}_X]). \tag{7.1}$$

One deduces from **SRR 1** and **SRR 2** a **Hirzebruch Riemann-Roch formula**

$$\chi(X, \mathscr{E}) = \int_X \operatorname{ch}(\mathscr{E}) \cap \operatorname{Td}(X). \tag{7.2}$$

Here $\int_X$ is the push-forward $f_{G.}$ for the morphism from X to S.

If X is a local complete intersection over S, then

$$\operatorname{Td}(X) = \operatorname{td}(T_{X/S}) \cap [\mathcal{O}_X]. \tag{7.3}$$

This formula (7.3) is a special case of a Riemann-Roch theorem which is dual to the Grothendieck Riemann-Roch theorem.

SRR 3. (Verdier Riemann-Roch). *If $f: X \to Y$ is a regular morphism, then the following diagram is commutative:*

$$\begin{array}{ccc} K.Y & \xrightarrow{\ \tau\ } & \mathbf{Q}G.Y \\ {\scriptstyle f^{K.}}\downarrow & & \downarrow{\scriptstyle f^{G.}} \\ K.X & \xrightarrow[\operatorname{td}(T_f)^{-1} \cap \tau]{} & \mathbf{Q}G.X \end{array}$$

Applying **SRR 3** to $Y = S$, $[\mathcal{O}_S] \in K.S$, yields (7.3).

The construction of τ may be sketched as follows. Given a coherent sheaf $\mathscr{F}$ on X, one imbeds X in a scheme P which is smooth over S, and one resolves $\mathscr{F}$ by a complex $\mathscr{E}^{\cdot}$ of locally free sheaves on P. Since $\mathscr{E}^{\cdot}$ is exact on $P - X$, the class

$$\sum (-1)^i \operatorname{ch}(\mathscr{E}^i) \cap [\mathcal{O}_P] \in \mathbf{Q}G.P$$

restricts to zero on $P - X$. Proposition 6.5 motivates the existence of a class in $\mathbf{Q}G.X$ whose image in $\mathbf{Q}G.P$ is this class. An essential step is to construct a *canonical* such class

$$\operatorname{ch}_X^P(\mathscr{E}^{\cdot}) \in \mathbf{Q}G.X.$$

Its construction is based on MacPherson's graph construction, which is a generalization of the deformation to the normal bundle. Then one defines

$$\tau_X([\mathscr{F}]) = \operatorname{td}(T_{P/S}) \cap \operatorname{ch}_X^P(\mathscr{E}.).$$

We refer to [BFM 1], [V], and [F] for details on MacPherson's graph construction, and the proof that τ_X is well defined and satisfies **SRR 1**, **SRR 2**, **SRR 3**, as well as for applications in the case of algebraic schemes over a field.

There is an entirely parallel discussion for the Adams operators

$$\psi^j: K^{\cdot} \to K^{\cdot}$$

One constructs

$$\psi_j: K.X \to \mathbf{Z}[1/j] \otimes K.X$$

satisfying the analogues of **SRR 1**, **SRR 2**, **SRR 3**. As in Chapter V, §7, $\theta^j(T_f^{\vee})^{-1}$ replaces the Todd classes $\mathrm{td}(T_f)$. The construction of $\psi_j([\mathscr{F}])$ is also analogous to that of τ, by imbedding X in a smooth P and resolving $\mathscr{F}$ by a complex of locally free sheaves; it would be interesting to find a more direct description of $\psi_j[\mathscr{F}]$. For an extension to higher K-theory, which follows the same pattern, see Soulé [S].

Remark. It follows easily from Proposition 5.1 that there is a functorial surjective homomorphism

$$A_m(X) \to G_m(X)$$

from the group of m-cycles module rational equivalence to the associated graded group to $K.(X)$. In fact, the Riemann–Roch theorem is proved with values in $\mathbf{Q}A_m(X)$, from which it follows that the above homomorphism becomes an isomorphism after tensoring with $\mathbf{Q}$. For details and more on rational equivalence, see [F 2].

VI §8. The Complex Case

For schemes over $S = \mathrm{Spec}(\mathbf{C})$, one has topological functors, the (singular, even) cohomology

$$H^{\cdot}(X) = \bigoplus_i H^{2i}(X; \mathbf{Z})$$

and

$$K^{\cdot}_{\mathrm{top}}(X),$$

the Grothendieck group of topological vector bundles on X. These are contravariant, ring-valued functors. Since vector bundles on X have Chern classes in $H^{\cdot}X$, there is a Chern character, which we denote

$$\mathrm{ch}^{\cdot}: K^{\cdot}_{\mathrm{top}} \to \mathbf{Q}H^{\cdot}$$

as in Chapter II, which is a natural transformation of contravariant functors; here $\mathbf{Q}H^{\cdot}$ denotes $H^{\cdot}(\ ;\mathbf{Q})$.

If $f: X \to Y$ is a projective local complete intersection morphism, there are push-forward homomorphisms

$$f_{H^{\cdot}}: H^{\cdot}X \to H^{\cdot}Y \qquad \text{and} \qquad f_{K^{\cdot}_{\mathrm{top}}}: K^{\cdot}_{\mathrm{top}}X \to K^{\cdot}_{\mathrm{top}}Y,$$

so that the diagram

$$\begin{array}{ccc} K^{\cdot}_{\mathrm{top}} & \xrightarrow{\mathrm{td}(T_f)\cdot \mathrm{ch}^{\cdot}} & \mathbf{Q}H^{\cdot}X \\ \downarrow{\scriptstyle f_{K^{\cdot}_{\mathrm{top}}}} & & \downarrow{\scriptstyle f_{H^{\cdot}}} \\ K^{\cdot}_{\mathrm{top}} & \xrightarrow{\mathrm{ch}^{\cdot}} & \mathbf{Q}H^{\cdot}Y \end{array}$$

commutes, i.e. Riemann–Roch holds for f with respect to $(K^{\cdot}_{\mathrm{top}}, \mathrm{ch}^{\cdot}, \mathbf{Q}H^{\cdot})$, with multiplier $\mathrm{td}(T_f)$. For the constructions, see [BFM 1] and [BFM 2].

There is a homomorphism

$$\alpha^{\cdot} = K^{\cdot}X \to K^{\cdot}_{\mathrm{top}}X$$

which takes $\mathscr{E}$ to $\mathbf{V}(\mathscr{E}^{\vee})$, where $\mathbf{V}(\mathscr{E}^{\vee})$ is the vector bundle whose sheaf of sections is $\mathscr{E}$ (Chapter IV, §1). This $\alpha^{\cdot}$ gives a natural transformation of contravariant functors. If $f: X \to Y$ is a projective local complete intersection morphism then the diagram

$$\begin{array}{ccc} K^{\cdot}X & \xrightarrow{\alpha^{\cdot}} & K^{\cdot}_{\mathrm{top}}X \\ \downarrow{\scriptstyle f_{K^{\cdot}}} & & \downarrow{\scriptstyle f_{K^{\cdot}_{\mathrm{top}}}} \\ K^{\cdot}Y & \xrightarrow{\alpha^{\cdot}} & K^{\cdot}_{\mathrm{top}}Y \end{array}$$

commutes, i.e. Riemann–Roch holds for f with respect to $(K^{\cdot}, \alpha^{\cdot}, K^{\cdot}_{\mathrm{top}})$ (with multiplier 1!). It follows (Chapter II, Proposition 1.4) that Riemann–Roch holds for the composite functor, i.e.,

$$\begin{array}{ccc} K^{\cdot}X & \xrightarrow{\mathrm{td}(T_f)\cdot \mathrm{ch}^{\cdot}\circ\alpha^{\cdot}} & \mathbf{Q}H^{\cdot}X \\ \downarrow{\scriptstyle f_{K^{\cdot}}} & & \downarrow{\scriptstyle f_{H^{\cdot}}} \\ K^{\cdot}Y & \xrightarrow{\mathrm{ch}^{\cdot}\circ\alpha^{\cdot}} & \mathbf{Q}H^{\cdot}Y \end{array}$$

commutes.

Once the push-forward homomorphisms are constructed for $K^{\cdot}_{\text{top}}$ and $H^{\cdot}$, these Riemann–Roch theorems may be proved by exactly the same procedure as in this treatise: by deformation to the normal bundle, the general case is reduced to the case of elementary imbeddings and projections.

In topology cohomology theories $H^{\cdot}$ and $K^{\cdot}_{\text{top}}$ are dual to homology theories $H_{.}$ and $K^{\text{top}}_{.}$, $\text{ch}^{\cdot}$ corresponds to a natural transformation

$$\text{ch}_{.}\colon K^{\text{top}}_{.} \to H_{.}$$

satisfying analogues of **SRR 1**, **SRR 2**. With $K_{.}$ as in §3, one can construct

$$\alpha_{.}\colon K_{.} \to K^{\text{top}}_{.}$$

"dual" to $\alpha^{\cdot}$, satisfying analogues of **SRR 1**, **SRR 2**, **SRR 3**. The proof follows the same pattern (cf. [BFM 2]).

A more intrinsic construction of $\alpha_{.}$, valid for arbitrary complex analytic spaces, and extending to higher K-theory, has recently been given by R. Levy.

VI §9. Lefschetz Riemann–Roch

The formalism developed here can also be used in another situation, to study *equivariant* sheaves.

Let k be an algebraically closed field, $S = \text{Spec}(k)$, and let n be a positive integer not divisible by the characteristic of k. Let $\mathfrak{C}$ be the category whose objects are pairs (X, h_X), where X is a smooth projective scheme over S, and

$$h_X : X \to X$$

is an endomorphism such that $h_X^n = \text{id}_X$. A **morphism**

$$f\colon (X, h_X) \to (Y, h_Y)$$

is a morphism $f\colon X \to Y$ such that $h_Y \circ f = f \circ h_X$. The hypotheses imply that the fixed point scheme of h_X on X, denoted X^h, is also smooth over S. A morphism f as above induces a morphism

$$f^h\colon X^h \to Y^h$$

of the fixed point schemes.

An **equivariant** (locally free) sheaf on (X, h_X) is a (locally free) sheaf $\mathscr{E}$ on X together with a homomorphism

$$\varphi_E \colon h_X^* \mathscr{E} \to \mathscr{E}.$$

(Note that it is not required that φ_E have finite order.) Homomorphisms and exact sequences of equivariant sheaves are defined in an evident way, so that one has a Grothendieck group $K(X, h_X)$ of equivariant locally free sheaves.

If h_X acts trivially on X, then

$$K(X, h_X) = K(X) \otimes_{\mathbf{Z}} \mathbf{Z}[k],$$

where $\mathbf{Z}[k]$ is the free abelian group on the elements of k, a ring with multiplication induced by multiplication in k. This is because any equivariant $\mathscr{E}$ is a finite sum of sheaves $\mathscr{E}_a$, for eigenvalues $a \in k$, on which $\varphi_a - a$ is nilpotent.

Fix a $\mathbf{Z}[k]$-algebra Λ such that for every n-th root of unity a in k, $a \neq 1$, the image of $[1] - [a]$ in Λ is invertible. (Note that such Λ can have characteristic zero, even if k has positive characteristic; e.g. Λ may be a Witt ring.)

For any (X, h_X) in $\mathfrak{C}$, the conormal sheaf $\mathscr{C} = \mathscr{C}_{X^h/X}$ to X^h in X is an equivariant sheaf on X^h, all of whose eigenvalues are non-trivial roots of unity. It follows that the element

$$\lambda_X = \sum (-1)^i [\Lambda^i \mathscr{C}] \in K(X^h) \otimes_{\mathbf{Z}} \Lambda$$

is invertible.

The functor $(X, h_X) \mapsto K(X, h_X)$ is both contravariant and covariant on $\mathfrak{C}$, just as in the absolute case. For this one needs to know that any equivariant coherent sheaf is the image of an equivariant locally free sheaf; this follows from the fact that any (X, h_X) admits a closed imbedding into (P, h_P) where P is a projective space over k, and h_P is a linear endomorphism.

An equivariant locally, free sheaf on (X, h_X) restricts to an equivariant locally free sheaf on (X^h, id), giving rise to a homomorphism

$$\rho \colon K(X, h_X) \to K(X^h) \otimes \Lambda$$

Thus if one defines $L(X, h_X) = K(X^h) \otimes \Lambda$, then (K, ρ, L) is a Riemann-Roch functor in the sense of Chapter II, §1.

The **Lefschetz Riemann–Roch theorem** asserts that for a morphism $f\colon (X, h_X) \to (Y, h_Y)$, the diagram

$$\begin{array}{ccc} K(X, h_X) & \xrightarrow{\lambda_f \cdot \rho} & K(X^h) \otimes \Lambda \\ \downarrow f_K & & \downarrow f_L \\ K(Y, h_Y) & \xrightarrow{\rho} & K(Y^h) \otimes \Lambda \end{array}$$

commutes, where

$$\lambda_f = \lambda_X^{-1} \cdot f^L(\lambda_Y) \in K(X^h) \otimes \Lambda.$$

The reader should find the proof a pleasant exercise: One factors f into a closed imbedding followed by a projection. The case of an imbedding is handled by deforming to the normal bundle, and calculating directly for an elementary imbedding. For a projection one proves equivariant analogues of the results of Chapter V, §2; the calculations for a projection are easiest for one of the form

$$(Y, h_Y) \times_S (P, h_P) \to (Y, h_Y),$$

with (P, h_P) as above. For details, as well as generalizations to the singular case, see [BFQ].

As a special case, one has a **fixed-point formula**. If the fixed point set X^h is finite, then

$$\sum_i (-1)^i \operatorname{tr}(H^i(X, \mathscr{E})) = \sum_{P \in X^h} \operatorname{tr}(\mathscr{E}(P))/\det(I - t_P).$$

Here for an equivariant vector space V over k, $\operatorname{tr}(V)$ is its image in Λ under the canonical homomorphism

$$K(S, \mathrm{id}) = \mathbf{Z}[k] \to \Lambda,$$

$\mathscr{E}(P)$ is the fibre (restriction) of $\mathscr{E}$ at P, and

$$t_P\colon T_P^*X \to T_P^*X$$

is the map on the cotangent space $T_P^*X = \mathscr{C}_{P/X}$ induced by h_X.

By Proposition 1.4 of Chapter II, this Lefschetz Riemann–Roch theorem can be composed with Grothendieck Riemann–Roch, yielding a commutative diagram

$$\begin{array}{ccc} K(X, h_X) & \xrightarrow{\tau_f \cdot \mathrm{ch} \circ \rho} & \mathbf{Q}\mathrm{Gr}\, K(X^h) \otimes \Lambda \\ \Big\downarrow{\scriptstyle f_K} & & \Big\downarrow{\scriptstyle f^h_{\mathbf{Q}\mathrm{Gr}\,K}} \\ K(Y, h_Y) & \xrightarrow{\mathrm{ch} \circ \rho} & \mathbf{Q}\mathrm{Gr}\, K(Y^h) \otimes \Lambda \end{array}$$

where

$$\tau_f = \mathrm{td}(T_{f^h}) \cdot \mathrm{ch}(\lambda_f).$$

Or one may compose with Adams Riemann–Roch....

As a final exercise, the reader may work out the analogous theorem when $\mathfrak{C}$ is replaced by the category of smooth projective schemes X over a finite field $k = \mathbf{F}_q$, and X^h is the set of $\mathbf{F}_q$-valued points of X (the fixed points of the Frobenius on X). Let $K'(X)$ be the Grothendieck group of locally free sheaves $\mathscr{E}$ on X together with q-linear endomorphism,

$$\varphi_{\mathscr{E}}: \mathscr{E} \to \mathscr{E}$$

(i.e. $\varphi_{\mathscr{E}}$ is additive, and $\varphi_{\mathscr{E}}(as) = a^q \varphi_{\mathscr{E}}(s)$ for a and s sections of $\mathscr{O}_X$ and $\mathscr{E}$ over an open set of X). When $X = \mathrm{Spec}(\mathbf{F}_q)$, such $\mathscr{E}$ is just a vector space with a linear map, and

$$K'(\mathrm{Spec}(\mathbf{F}_q)) \xrightarrow[\mathrm{tr}]{\approx} \mathbf{F}_q.$$

Therefore for any X, $K'(X^h)$ is a vector space over $\mathbf{F}_q$ with basis the points in X^h. Restriction gives a Riemann–Roch functor

$$\rho: K'(X) \to K'(X^h).$$

For $f: X \to Y$, one has a **Frobenius Riemann–Roch theorem**: the diagram

$$\begin{array}{ccc} K'X & \xrightarrow{\rho} & K'(X^h) \\ \Big\downarrow{\scriptstyle f_{K'}} & & \Big\downarrow{\scriptstyle f^h_{K'}} \\ K'Y & \xrightarrow{\rho} & K'(Y^h) \end{array}$$

commutes. In particular, given $\mathscr{E}$ on X, we have the formula

$$\sum_i (-1)^i \operatorname{tr}(H^i(X, \mathscr{E})) = \sum_{P \in X^h} \operatorname{tr}(\mathscr{E}(P)).$$

For example, if $H^i(X, \mathcal{O}_X) = 0$ for $i > 0$, and X is geometrically connected, then

$$\# X(\mathbf{F}_q) \equiv 1 \mod p,$$

where q is a power of the prime p, a Chevalley–Warning formula. For details and a generalization to the singular case, see [F 1].

References

[EGA] A. GROTHENDIECK, with J. DIEUDONNÉ, *Éléments de géometrie algébrique*, Publ. Math. I.H.E.S. **4**, **8**, **11**, **17**, **20**, **24**, **28**, **32** (1961-1967)

[SGA 6] P. BERTHELOT, A. GROTHENDIECK, L. ILLUSIE, *et al.*, *Théorie des intersections et théorème de Riemann-Roch*, Springer Lecture Notes 225, 1971

[AK] A. ALTMAN, S. KLEIMAN, *Introduction to Grothendieck duality theory*, Springer Lecture Notes 146, 1970

[At] M. ATIYAH, *K-Theory*, Benjamin, 1967

[At-Hi] M. ATIYAH and F. HIRZEBRUCH, *Cohomologie-Operationen und charakteristische Klassen*, Math. Z. **77** (1961) pp. 149–187

[AT] M. F. ATIYAH, D. O. TALL, *Group representations, λ-rings and the J-homomorphism*, Topology **8** (1969) pp. 253–297

[BFM 1] P. BAUM, W. FULTON, R. MACPHERSON, *Riemann–Roch for singular varieties*, Publ. Math. I.H.E.S. **45** (1975) pp. 101–145

[BFM 2] ———, *Riemann–Roch and topological K-theory for singular varieties*, Acta. Math. **143** (1979) pp. 155–192

[BFQ] P. BAUM, W. FULTON, G. QUART, *Lefschetz–Riemann–Roch for singular varieties*, Acta. Math. **143** (1979) pp. 193–211

[BS] A. BOREL, J.-P. SERRE, *Le théorème de Riemann–Roch (d'apres Grothendieck)*, Bull. Soc. Math. France **86** (1958) pp. 97–136

[B] M. BORELLI, *Some results on ampleness and divisorial schemes*, Pacific J. Math. **23** (1967) pp. 217–227

[Bo] R. BOTT, *Lectures on* $K(X)$, Benjamin, 1969

[Ev] L. EVENS, *On the Chern classes of representations of finite groups*, Trans. Amer. Math. Soc. **115** (1965) pp, 180–193

[Ev-K] L. EVENS and D. S. KAHN, *An integral Riemann–Roch formula for induced representations for finite groups*, Trans. Am. Math. Soc. **245** (1978) pp. 809–330

[F 1] W. FULTON, *A fixed point formula for varieties over finite fields*, Math. Scand. **42** (1978) pp. 189–196

[F 2] W. FULTON, *Intersection Theory*, Springer-Verlag, 1984

[FM] W. FULTON, R. MACPHERSON, *Categorical framework for the study of singular spaces*, Mem. Amer. Math. Soc. **243**, 1981

[Gr] A. GROTHENDIECK, *Classes de Chern et représentations linéaires des groupes discrets*, Dix éxposés sur la cohomologie étale des schemas, North-Holland, Amsterdam, 1968

[H] R. HARTSHORNE, *Algebraic geometry*, Springer-Verlag, 1977

[Hi] F. HIRZEBRUCH, *Neue topologische Methoden in der algebraischen Geometrie*, Ergebnisse der Mathematik, Springer-Verlag, 1956; Translated and expanded to the English edition, *Topological Methods in Algebraic Geometry*, Grundlehren der Mathematik, Springer-Verlag, 1966

[J] J. P. JOUANOLOU, *Riemann-Roch sans dénominateurs*, Inv. Math. **11**, (1970) pp. 15–26

[Ke] M. KERVAIRE, *Opérations d'Adams en théorie des représentations linéares des groupes finis*, l'Ens. Math. **22** (1976) pp. 1–28

[Kn] J. KNOPFMACHER, *Chern classes of representations of finite groups*, J. London Math. Soc. **41** (1965) pp. 535–541

[Kn] D. KNUTSON, *λ-rings and the representation theory of the symmetric group*, Springer Lecture Notes 308, 1973

[Kr] Ch. KRATZER, *Opérations d'Adams et représentations de groupes*, l'Ens. Math. **26** (1980) pp. 141–154

[L] S. LANG, *Algebra*, second edition, Addison-Wesley, 1984

[Man] Y. I. MANIN, *Lectures on the K-functor in algebraic geometry*, Russ. Math. Surveys **24**, No. 5 (1969) pp. 1–89

[Mat] H. MATSUMURA, *Commutative algebra*, second edition, Benjamin/Cummings, 1980

[Mi] A. MICALI, *Sur les algèbres universelles*, Ann. Inst. Fourier, Grenoble **14** (1964) pp. 33–88

[Q] D. QUILLEN, *Higher algebraic K-theory*: I, Springer Lecture Notes 341, 1973 pp. 85–147

[S] C. SOULÉ, *Opérations en K-theorie algébrique*, CNRS preprint, 1983

[Th 1] C. B. THOMAS, *Riemann–Roch formulae for group representations*, Mathematika **20** (1973) pp. 253–262

[Th 2] C. B. THOMAS, *An integral Riemann–Roch formula for flat line bundles*, Proc. London Math. Soc. **XXXIV** (1977) pp. 87–101

[V] J.-L. VERDIER, *Le théorème de Riemann–Roch pour les intersections complètes*, Astérisque **36–37** (1976) pp. 189–228

Index of Notations

Index

Grundlehren der mathematischen Wissenschaften

Continued from page ii

235. Dynkin/Yushkevich: Markov Control Processes and Their Applications
236. Grauert/Remmert: Theory of Stein Spaces
237. Köthe: Topological Vector-Spaces II
238. Graham/McGehee: Essays in Commutative Harmonic Analysis
239. Elliott: Probabilistic Number Theory I
240. Elliott: Probabilistic Number Theory II
241. Rudin: Function Theory in the Unit Ball of C^n
242. Huppert/Blackburn: Finite Groups I
243. Huppert/Blackburn: Finite Groups II
244. Kubert/Lang: Modular Units
245. Cornfeld/Fomin/Sinai: Ergodic Theory
246. Naimark/Štern: Theory of Group Representations
247. Suzuki: Group Theory I
249. Chung: Lectures from Markov Processes to Brownian Motion
250. Arnold: Geometrical Methods in the Theory of Ordinary Differential Equations
251. Chow/Hale: Methods of Bifurcation Theory
252. Aubin: Nonlinear Analysis on Manifolds, Monge-Ampère Equations
253. Dwork: Lectures on p-adic Differential Equations
254. Freitag: Siegelsche Modulfunktionen
255. Lang: Complex Multiplication
256. Hörmander: The Analysis of Linear Partial Differential Operators I
257. Hörmander: The Analysis of Linear Partial Differential Operators II
258. Smoller: Shock Waves and Reaction-Diffusion Equations
259. Duren: Univalent Functions
260. Freidlin/Wentzell: Random Perturbations of Dynamical Systems
261. Remmert/Bosch/Güntzer: Non Archimedian Analysis—A Systematic Approach to Rigid Analytic Geometry
262. Doob: Classical Potential Theory & Its Probabilistic Counterpart
263. Krasnosel'skiĭ/Zabreĭko: Geometrical Methods of Nonlinear Analysis
264. Aubin/Cellina: Differential Inclusions
265. Grauert/Remmert: Coherent Analytic Sheaves
266. de Rham: Differentiable Manifolds
267. Arbarello/Cornalba/Griffiths/Harris: Geometry of Algebraic Curves, Vol. I
268. Arbarello/Cornalba/Griffiths/Harris: Geometry of Algebraic Curves, Vol. II
269. Schapira: Microdifferential Systems in the Complex Domain
270. Scharlau: Quadratic and Hermitian Forms
271. Ellis: Entropy, Large Deviations, and Statistical Mechanics
272. Elliott: Arithmetic Functions and Integer Products
273. Nikolskij: Treatise on Shift Operators
274. Hörmander: The Analysis of Linear Partial Differential Operators III
275. Hörmander: The Analysis of Linear Partial Differential Operators IV
276. Liggett: Interacting Particle Systems
277. Fulton/Lang: Riemann-Roch Algebra
278. Barr/Wells: Toposes, Triples, and Theories
279. Bishop/Bridges: Constructive Analysis